Advanced Research Methods in the Built Environment

Advanced Research Methods in the Built Environment

EDITED BY

Andrew Knight
School of Architecture
Design and the Built Environment
Nottingham Trent University

Les Ruddock
School of the Built Environment
University of Salford

WILEY-BLACKWELL
A John Wiley & Sons, Ltd., Publication

This edition first published 2008
© 2008 Blackwell Publishing Ltd

Blackwell Publishing was acquired by John Wiley & Sons in February 2007. Blackwell's publishing programme has been merged with Wiley's global Scientific, Technical, and Medical business to form Wiley-Blackwell.

Registered office
John Wiley & Sons Ltd, The Atrium, Southern Gate, Chichester, West Sussex, PO19 8SQ, United Kingdom

Editorial offices
9600 Garsington Road, Oxford, OX4 2DQ, United Kingdom
2121 State Avenue, Ames, Iowa 50014-8300, USA

For details of our global editorial offices, for customer services and for information about how to apply for permission to reuse the copyright material in this book please see our website at www.wiley.com/wiley-blackwell.

The right of the author to be identified as the author of this work has been asserted in accordance with the Copyright, Designs and Patents Act 1988.

Library of Congress Cataloging-in-Publication Data

Advanced research methods in the built environment /edited by Andrew Knight, Les Ruddock.
    p. cm.
  Includes bibliographical references and index.
  ISBN 978-1-4051-6110-7 (pbk. : alk. paper)
  1. Building–Research. 2. Research–Methodology. I. Knight, Andrew,
  1972- II. Ruddock, Leslie, 1950-

TH213.5.A38 2008
690.072–dc22

                                                2008013079

A catalogue record for this book is available from the British Library.

Set in 9.5/11.5 pt Avenir by Newgen Imaging Systems Pvt. Ltd, Chennai, India

1   2008

Contents

Foreword

This book sets out to complement the more standard research methods textbooks available, by broadening and deepening the treatment given. A range of very experienced researchers thus provide perspectives on a wide variety of research paradigms, but there are also contributions concerning the 'nitty gritty' of research practice. This is all delivered solidly within the context of built environment research.

Together the contributions provide a wealth of wisdom and insights for the postgraduate researcher, or indeed the ambitious undergraduate or curious established researcher.

The diversity of the subjects covered is an indication of the complexity of the built environment research domain. The quality of the material is a very positive measure of the level of maturity that this research discipline has now reached.

Professor Peter Barrett MSc, PhD, DSc, FRICS
Pro-Vice Chancellor for Research, University of Salford
President of the CIB (International Council for Research and
Innovation in Building and Construction)

Introduction

The built environment is a diverse field attracting a wide variety of researchers approaching their object of study from different disciplinary and methodological perspectives. For a new researcher, this diversity can add both interest and challenge to the practice of research. Many research questions, which reside in the built environment, may require investigation using the theories and methodological tools from disciplines such as art, economics, law, philosophy, sociology and statistics, to name just a few. However, many built environment researchers have been academically trained in professional areas; for example, architecture, construction management, engineering or surveying. Even though most of these degrees require the undertaking of research projects to some extent, and many include tuition in subjects such as applied statistics and economics, the principal focus of curriculum is preparation for professional practice. As such, a transition to postgraduate research work (and beyond) can be bewildering for the student. The principal aim of this book is to provide a bridge between the introductory research methods books typically used at undergraduate level and the many discipline-specific texts, which can be difficult to access for the non-specialist.

In a book of this nature, it would be impossible to comprehensively cover all areas of interest to researchers in the built environment. Instead, our aim has been to provide a text covering a variety of topics which typically concern new researchers. The selection of topics has been informed through consultation with the specialist chapter authors, all of whom have years of experience as both researchers and supervisors. Topics range from pragmatic issues surrounding the production of a thesis or journal article, to chapters considering the role of theory or epistemology in research. Our intention is that each chapter is the start of a journey, allowing readers an opportunity to develop a familiarity with an area before deciding whether to progress to more specialised texts.

Although the primary audience for this book is the postgraduate, enthusiastic undergraduates will also find various chapters interest. Additionally, owing to the diversity of our field, academics will also find some of the specialist chapters useful. Therefore, this is not the sort of book to be formulaically read cover to cover. It is essentially a resource book covering a wide variety of issues ranging from the philosophical to the purely practical. Hence, to some degree, we feel the necessary constraint of the concise book title may be too narrow to do justice to its broad content.

Like choosing a title, the order and subjects of the chapters posed another challenge. Ultimately, we believed that sectioning the book into various chapter groups would provide a false impression of clear subject divisions. For example, many chapters touch on data collection as well as broader methodological issues. However, there is a logic to the order. We start with broad subject areas of enquiry, before turning to more generic concerns. The book then moves into examples of some very specific methods before the final part focuses on more pragmatic issues. A brief overview and discussion of some of the connectivity and complementarity between the chapters is provided below.

Chapter One opens the book by calling for methodological pluralism and, as such, provides a useful starting point. Dainty argues that, despite philosophical debates in the field of construction management in the 1990s, there has been reluctance on the part of researchers to fully embrace alternatives to the dominant quantitative paradigm. Drawing on an analysis of the methods used for papers published in one of the leading journals, it appears that the construction management research community is still firmly rooted in the positivist tradition. The chapter concludes by arguing that if researchers are to get a fuller understanding of why those in the construction process do what they do, a more holistic and adventurous approach to research is required.

In Chapter Two, the role of architectural research in the built environment is considered. Penn argues that architectural research is important to understanding the production and management of the built environment. However, he goes further in this chapter by arguing that architectural research is also crucial to architectural practice. One important area of research is then examined: the analysis of spatial layout in plan. The chapter concludes by highlighting the tensions between two kinds of knowledge: scientific and social. Finally, the drift towards the objectified, explicit and scientific knowledge is discussed and the limitations of this approach are raised. Parallels with Dainty's arguments for methodological pluralism in construction management are clear.

A third area of enquiry in the built environment is examined in Chapter Three – that of legal research. Here, Chynoweth recognises the array of component disciplines that comprise the built environment and states the importance of increased understanding across, what can be conceptualised as, an interdisciplinary or multidisciplinary field. The specific nature of the epistemological, methodological and cultural features of legal research is then explained. In common with the first two chapters, Chynoweth concludes with comments on the dominance of the scientific approach to knowledge acquisition in the built environment. He argues that, for legal researchers, this can cause problems when communicating their work, since empirical investigation is not their primary concern.

Morton and Wilkinson state that the built environment professions are male-dominated. Hence, gender-related research to understand the field, and to bring about change, is both necessary and desirable. It is argued that feminist research is conducted for, by and about women, and the authors explain that, despite higher numbers of women entering higher education in the 1990s, discrimination and barriers to progression are still experienced. They conclude by summarising various approaches to feminist research with examples of work undertaken in the field.

As argued by Chynoweth, the built environment is multidisciplinary. One important discipline for many researchers in our field is economics. However, in many respects the construction industry is unlike the generalised industry of economic theory. In Chapter Five, Ruddock discusses issues involved in undertaking economic analysis and attempting to model the built environment sector, in the context of its relationship with the wider economy. Research in the field of construction economics often means applied research in the field to test the validity of hypotheses. This requires meaningful analysis of the data surrounding the sector, and a feature of this chapter is consideration of how data are used to analyse this relationship with the rest of the economy.

A constant thread of interest through many of the chapters in this book is the issue of knowledge. In this sense, philosophy is not another discipline within the arena of built environment research, but a foundation. Questions concerning the process of knowledge acquisition and its limits are fundamental to methodology in all areas whether legal, technological, social or economic. Hence, Chapter Six explores the subject known as epistemology. Knight and Turnbull commence by arguing that many postgraduate researchers struggle when they first start to consider issues surrounding knowledge in their discipline. One reaction is to ignore the problem; another reaction is to get so engrossed in reading philosophy books that practical progress is jeopardised. The aim of Chapter Six is to provide a crossing for the non-specialist into this challenging terrain. It is argued that, since many words such as 'postmodernism' and 'positivism' are used in an inconsistent manner, the most appropriate way to engage in epistemology is through an historical appreciation of the development of key ideas and thinkers. Hence, the chapter takes the reader on a journey from classical to modern epistemology and concludes with a reflection on more recent 'postmodern' ideas.

Many of the chapters in this book advocate more pluralism and, in particular, a movement towards what can broadly be described as qualitative methods. However, what may be seen as a counter-argument in support of scientific theory is forwarded by Runeson and Skitmore. What makes this chapter particularly interesting is that these authors were at the vanguard of the mid-1990s debate referred to by Dainty in Chapter One. Chapter Seven explores the logic of scientific theory and the importance of working as a scientist. The arguments throughout this chapter are essentially Popperian and this provides interesting reading for those undertaking experimental work. However, those who may be attempting to justify an alternative to a scientific approach should also consider reflecting on the authors' arguments.

A popular theoretical alternative to the scientific approach, which is often but not exclusively associated with qualitative methods, is grounded theory. This methodology allows researchers to work inductively, from data to theory, and is often used as a model in projects, in which there is a lack of relevant testable theory. In Chapter Eight, Hunter and Kelly explore the literature surrounding grounded theory. They draw out a number of issues concerning the role of literature in theory development and data collection strategies. One important area for qualitative researchers to consider is the range of analytical strategies available. In this chapter, data coding is specifically considered and this will provide a foundation for those interested in computer-based qualitative analysis, as discussed by King in Chapter Twelve.

Many readers will be aware of case study research, even though they may have difficulty defining what a case study actually is. The confusion around the nature of this methodology is acknowledged by Proverbs and Gameson in Chapter Nine. The chapter considers a variety of issues around designing, identifying and selecting the cases. When undertaking case studies, researchers are often overwhelmed by the information collected so an organised approach is required. The authors outline some of the data options available including: documents, archival records, interviews, detached or direct observations, participant observation and physical artefacts. A practical example of a case study on the topic of project team dynamics is used to demonstrate the use of this methodology in a built environment context.

As the reader progresses through the book, it should be clear that there is a movement from theory to practice and it is the practice of undertaking interviews which is explored by Haigh. Interviews are used widely by researchers and vary in type and structure, making them an extremely versatile method. However, as argued in this chapter, this is not a clinical objective data collection tool in most situations. The human dynamics between researcher and researched complicate the picture and this relationship is at the core of understanding the nature of interviewing.

As an alternative, or in addition to interviews, many projects use a questionnaire, particularly when quantitative data are required. In Chapter Eleven, Hoxley examines the design and use of questionnaires. He considers the importance of wording, structure, sampling and coding. The second part of the chapter examines factor analysis – a popular technique in statistics for data reduction. Attempting to reduce many indicator variables down to a few underlying concepts is the aim of this technique. Using a practical example of the service quality experienced by clients of professional surveyors, Hoxley demonstrates the procedure using the computer-based statistical package SPSS.

The next chapter also examines the use of computers in data analysis, but this time from the perspective of analysing qualitative data. In Chapter Twelve, King outlines the important benefits that computers bring to research analysis, particularly when dealing with a large number of codes. However, he also argues that there are many dangers for the novice researcher, especially when the software starts to dominate the analysis. Additional problems surrounding training are also considered. Chapter Thirteen is again focused on data analysis, but this time from a quantitative perspective. In this chapter, Leishman introduces the reader to a range of statistical concepts and tests that may be useful to researchers working in the built environment. The reader is provided with sufficient background to begin exploring issues in the application of statistical techniques.

Occasionally, when new researchers read about techniques being used by academics, it can be difficult to gain an understanding of the method. This may be for two reasons. Firstly, in journal articles, authors have to remain within strict word limits and focus on the outcomes of the research rather than the methods. Secondly, many textbooks that cover such techniques can be highly specialised and difficult to access for anyone new to the area. In Chapters Fourteen and Fifteen, two such techniques are described where the authors are given the space to focus on method rather than the outcomes of their research. In the first of these chapters, Boussabaine and Kirkham explore the use of artificial neural network (ANN) modelling. The authors provide the reader with a general introduction to ANNs, some background to methods and methodology, a review of current research and an evaluation of the most significant applications of ANN techniques. In the second, Pryke explains the use of social network analysis. He argues that since this method is capable of analysing qualitative concepts through a mathematical and graphical approach, there is enormous scope for more utilisation of this form of structural analysis in built environment research.

The final four chapters in this book turn to more generic issues of a practical nature. Griffith and Watson, in Chapter Sixteen, cover a topic almost all postgraduate researchers will be concerned by: managing the thesis. Here they explore a number of issues including questioning what a thesis actually is. This is a very important matter for research students since ultimately the thesis, and the associated oral examination, will form the basis of the final assessment. Throughout this chapter, the authors focus on the need to project manage the thesis, and how students should accept personal responsibility for this important task. Additionally, there is practical advice for those who need to prepare for an oral examination of their thesis.

For those considering an academic career, developing a portfolio of publications is critical. In Chapter Seventeen, Hughes (who currently edits one of the leading journals in the built environment with Dainty) reflects on the practice and process of publishing academic journal papers. This chapter is full of valuable advice, giving the reader an interesting insight into the views of an experienced editor. Hughes argues that with a clearer understanding of the processes involved, authors can significantly improve their success rate in getting papers published.

Nevertheless, sometimes things just do not go to plan. We can lose interest in our work long before it gets to the stage of publication or even thesis. This can happen for a variety of reasons. We may move to a new job, sometimes we just feel there are more important things in life than a research project, at other times personal factors outside our control change. However, having spent months or even years on a project, failure to complete a research degree is often something which may remain a lifelong regret. Knowing how to deal with the inevitable motivational lows, is probably one of the main factors which determine our ability to succeed. In Chapter Eighteen, Boyd tackles the issue of researcher fatigue. He offers some theoretical insights in to understanding ourselves, in addition to giving practical advice on how to keep going.

The final chapter looks to the future of built environment research. In this concluding chapter, Ratcliffe puts forward the idea that the methods of the past may not be the most profitable for stretching the boundaries of knowledge in the twenty-first century.

This, in many respects, takes us 'full circle' since this is the contention of several of the chapters at the start of this book. It is hoped that this book will contribute to a new sense of shared methodological understanding, both across and within the disciplinary fields that broadly constitute built environment research.

Andrew Knight and Les Ruddock
2008

Contributors

About the editors

Dr Andrew Knight
Andrew Knight is a Principal Lecturer in Construction Economics in the School of Architecture, Design and the Built Environment, Nottingham Trent University. Originally a surveyor, he now organises and teaches research methods across the University in addition to supervising research students. His research interests focus on the economics of construction, procurement and philosophy.

Professor Les Ruddock
Les Ruddock is Professor of Construction and Property Economics in the School of the Built Environment and Associate Dean for Research in the Faculty of Business, Law and the Built Environment at the University of Salford. He is Co-ordinator of the CIB Task Group on Macroeconomics for Construction and has written extensively on the economics of the construction and built environment sectors.

About the authors

Dr Abdelhalim Boussabaine
Abdelhalim Boussabaine is a Senior Lecturer at the School of Architecture, University of Liverpool. He has published widely on cost modelling for researchers and professionals and has researched extensively in the fields of artificial neural networks and fuzzy logic. His recent texts include *Whole Life Cycle Costing: Risk and Risk Responses* (2004) and *Cost Planning of PFI Projects* (2007).

Professor David Boyd
David Boyd is Professor of Construction and Deputy Head of the School of Property, Construction and Planning, at Birmingham City University. Although his background is in engineering, David is better known for his sociological and management research on the industry. He was chair of the Association of Researchers in Construction Management from 2006 to 2008 and organises a high profile think tank, the Next Steps Forum, chaired by Sir Michael Latham. He is co-author of *Understanding the Construction Client* (2006) published by Blackwell.

Paul Chynoweth
Paul Chynoweth is Senior Lecturer in Law in the School of the Built Environment, University of Salford. He is the editor of the *International Journal of Law in the Built Environment* and coordinator of CIB Working Commission (W113) on Law & Dispute

Resolution. He is a Solicitor of the Supreme Court of England and Wales and the author of *The Party Wall Casebook*, published by Blackwells.

Professor Andrew Dainty

Andrew Dainty is Professor of Construction Sociology in the Department of Civil and Building Engineering, Loughborough University. His research focuses on human social action within construction and other project-based sectors. He is co-author of *HRM in Construction Projects* (2003), *Communication in Construction* (2006) and is co-editor of *People and Culture in Construction* (2007), all published by Taylor & Francis.

Dr Rod Gameson

Rod Gameson is a Reader in Construction Management in the Construction and Infrastructure Department, University of Wolverhampton. He teaches research methods to postgraduate students, and also supervises a number of research students. His research, conducted in the UK and in Australia, focuses upon construction clients, project teams and their interactions.

Professor Alan Griffith

Alan Griffith is Professor of Construction Management at Sheffield Hallam University. His work has been published extensively and in recognition of his outstanding contribution to research and its dissemination he has received many literary and professional awards. He was awarded a higher doctorate in 2006.

Dr Richard Haigh

Richard Haigh is a Lecturer in Construction Management in the School of the Built Environment, University of Salford and has ten years of research experience. He teaches qualitative research methods to undergraduate and postgraduate students in the built environment disciplines and previously managed the School's research training programme.

Professor Mike Hoxley

Mike Hoxley is Professor of Building Surveying at Nottingham Trent University. He edits the Emerald journal *Structural Survey* and is the author of *A Construction Companion to Building Surveys* published by RIBA Publications. His research interests include managing and marketing the professional services firm, and building pathology.

Professor Will Hughes

Will Hughes is Professor of Construction Management and Economics and Head of the School of Construction Management and Engineering, University of Reading. He has an international reputation in the field of construction procurement and project organisation, focusing on commercial processes in construction procurement. He is Editor-in-chief of the international journal, *Construction Management and Economics*.

Dr Kirsty Hunter

Following completion of her PhD degree at Glasgow Caledonian University, Kirsty worked as Research Manager at Health Facilities Scotland and is currently working as Regional Project and Service Improvement Manager for the Health Protection Agency. She has worked as a PhD in value management and has worked on a variety of construction-related research projects. Through the dissemination of her research, she has won two best paper awards at international conferences and a highly commended Emerald journal award.

Professor John Kelly
John Kelly is Visiting Professor in the School of Architecture, Design and the Built Environment, Nottingham Trent University. Until recently he held the Morrison Chair in Construction Innovation at Glasgow Caledonian University. He has published numerous research papers and books in the area of value management and engineering.

Andrew King
Andrew King is Supply Chain Development Manager for Morgan Ashurst, one of the largest construction contractors in the UK. Before rejoining the industry, initially as a surveyor, he held a research position at Sheffield Hallam University where he published *Best Practice Tendering for Design and Build Projects* (2003) with Griffith and Knight.

Dr Richard Kirkham
Richard Kirkham is a Senior Lecturer in Construction Management and Quantity Surveying at the School of the Built Environment, Liverpool John Moores University. His research interests lie in whole life-cycle costing, advanced cost modelling techniques, the applications of stochastic processes to occupancy cost forecasting and service life prediction. He is author of *Ferry and Brandon's Cost Planning of Building and Whole Life-cycle Costing: Risk and Risk Responses*.

Dr Chris Leishman
Chris Leishman is a Senior Lecturer in Real Estate in the Department of Urban Studies, University of Glasgow. His research interests are focused on economic analysis and econometric modelling, particularly in the area of housing economics. He teaches development appraisal, urban and property economics and research methods.

Pat Morton
Pat Morton is a Principal Lecturer within the Centre for Science Education at Sheffield Hallam University. After 30 years in industry and teaching, she now leads a team working on a range of projects aimed at increasing the participation of women in the built environment and science, engineering and technology as well as working in research on gender and the built environment.

Professor Alan Penn
Alan Penn is Professor of Architectural and Urban Computing at The Bartlett, University College London, Director of Space Syntax Ltd., HEFCE Business Fellow and Chair of the Architecture & the Built Environment sub-panel for the UK Research Assessment Exercise 2008. His research investigates the way that architectural design affects the social and economic behaviour of organisations and communities.

Professor David Proverbs
David Proverbs is Professor of Construction Management and Head of the Construction and Infrastructure Department at the University of Wolverhampton. His research interests include productivity and performance issues in construction and climate change impacts on the built environment, specifically in the area of flooding.

Dr Stephen Pryke

Stephen Pryke is Director of Studies on the MSc Project and Enterprise Management programme at The Bartlett, University College London. He has carried out research with some of the largest construction organisations in the UK, France and China. Before entering academia he held a number of senior positions in industry, as well as running his own practice.

Professor John Ratcliffe

John Ratcliffe is Director and Dean of the Faculty of the Built Environment at the Dublin Institute of Technology, and Founder and Chairman of The Futures Academy there. Over the past decade he has acquired a particular expertise in the futures field and has just completed his terms as Secretary-General of the World Futures Studies Federation, the global body for professional futurists.

Dr Göran Runeson

Göran Runeson, recently retired, is Adjunct Professor at University of Technology, Sydney. After a few years in the Swedish merchant navy and construction industry, he obtained a first in Economics and worked as a lecturer in Economics and later Construction Management. He is the author of three books, 160 papers and supervisor of 20 past and present PhD students.

Professor Martin Skitmore

Martin Skitmore is currently Research Professor in the School of Urban Studies at Queensland University of Technology. He has authored or co-authored several books and over 100 academic journal papers, mainly concerned with price and cost modelling, contractor and consultant selection, and various aspects of project management. Martin is former CoEditor-in-Chief of the *Journal of Construction Innovation*.

Dr Neil Turnbull

Neil Turnbull is a Senior Lecturer in Philosophy and Social Theory at Nottingham Trent University in the UK. He is also one of the editor's of the journals *Theory, Culture & Society* and *Body and Society*.

Professor Paul Watson

Paul Watson in Professor of Building Engineering and Head of Construction, Cost and Surveying at Sheffield Hallam University. He has authored three books, and co-authored another three. Paul has also authored or co-authored 145 academic journal and conference papers. He has supervised and examined at PhD and DBA level, and his research interests are in the application of Quality Systems.

Sara Wilkinson

Sara Wilkinson is a Senior Lecturer in the Faculty of Architecture, Building and Planning at the University of Melbourne, Australia. She has published over 75 papers and research reports. Her current research investigates the adaptive reuse potential of existing buildings.

Chapter One
Methodological pluralism in construction management research

Andrew Dainty

Introduction

A fundamental question confronting anyone doing social research is for them to construct a philosophical position and orientation towards their enquiry. Unlike many domains which have established practices, construction management is a relatively new field which draws from both the natural and social sciences. As such, many different theories of knowledge or paradigms compete for methodological primacy. Researchers draw from both traditions when designing their research projects in a way which remains sensitive to the theoretical and philosophical foundations upon which their enquiry is based. However, the extent to which this has resulted in a plurality of methodological perspectives is questionable. For many years positivism and quantitative methods have been in the ascendancy in construction management research (Fellows and Liu, 1997: 78–79). This has promoted an orthodoxy of the application of 'natural science' methods to study social phenomena and an attendant focus on *explaining* human behaviour. In contrast, proponents of interpretivism, as an alternative paradigm, espouse the importance of *understanding* human behaviour (Bryman and Bell, 2003: 15). This has an emphasis on the empathetic comprehension of human action rather than the forces which shape it (ibid. 16). This perspective arguably has the potential to provide complementary insights, enriching understanding of the perspectives of those who work in the sector.

The construction management research community has an interesting history when it comes to debating the merits and demerits of different theoretical and philosophical perspectives on methodologies from different research paradigms. Concerns at the apparent dominance of positivism and the role of theory in construction management research in the mid-1990s led to a philosophical debate in the journal *Construction Management and Economics*. This debate was initiated by two papers in particular (Seymour and Rooke, 1995; Seymour *et al.*, 1997), which questioned the dominance of the rationalist position which seemingly underpinned most research within the community, suggesting that this tacitly endorsed the very attitudes in need of change in the industry. They suggested that the culture of research must change if researchers were to have an influence on the industry. In responding to Betts and Lansley's (1993) review of the first ten years of the Journal, Seymour *et al.* (1997) further questioned the dominance of the scientific theorising associated with realist ontological and epistemological positions, given that the 'object' of most construction management research is people. This suggested that the construction management discipline underestimated the interpretive process. These papers invoked a vigorous and somewhat polarised response around the relative merits of different research approaches.

Seymour and his colleagues were accused of being 'anti-scientific' and of propagating an approach which has yet to yield productive output, theories or progress (Runeson, 1997). Further, they were accused of promoting an approach more akin to consultancy than research, and of advocating methods which themselves have been widely criticised within the sociological literature (Harriss, 1998). Seymour and Rooke (1995) were also accused of setting out battle lines in the way that they dichotomised rationalist and interpretative paradigms to the detriment of research standards (Raftery et al., 1997). Seymour and colleagues defended their position by counter claiming that Raftery et al. themselves undermined standards by failing to recognise that different methods suit different purposes and that their position was symptomatic of the widespread confusion over terms such as 'method', 'methodology' and 'paradigm' (Rooke et al., 1997). They also questioned Runeson's definition of 'science', defending the rigour of the methods associated with the interpretive paradigm and their value in establishing the meaning ascribed by the social actors being studied (Seymour et al., 1998). Various other authors weighed into the debate (Loosemore et al., 1996), with some questioning its value given the apparent focus on research methods as opposed to methodology (Root et al., 1997).

More than a decade on, a number of questions emerge in terms of the legacy of this debate in terms of the impact it has had on construction management research. Firstly, have alternative research paradigms been embraced, or did the construction management community merely revert to its traditional adherence to positivism and quantitative methods? Secondly, do those within the construction management community draw upon a greater diversity of methods to enrich their understanding of the actuality of practice from the perspectives of those who work in the sector? And thirdly, has there been a move towards mixing paradigms and methods, or have the rival camps within the construction management research community remained entrenched and dichotomised within their own ontological and epistemological communities? This chapter aims to attempt to provide some answers to these questions in order to establish whether the debate has had a lasting legacy on the way in which construction management researchers now 'do' social research. In particular, it examines the extent to which construction management researchers have embraced 'multi-strategy' research – that which integrates quantitative and qualitative research within a single design (cf. Layder, 1993; cited in Bryman and Bell, 2003). In management science research, this perspective has been most recently associated with 'multimethodology', the practice of combining methodologies from different paradigms in an attempt to providing richer insights into relationships and their interconnectivities within organisations (Mingers and Gill, 1997). In advocating such a position, the aim is not to infer that combining strategies is inherently 'better' than employing a single research strategy, but to present an alternative perspective on how construction management researchers might design their research projects in the future.

Initially, the basic principles of research strategy and design are examined and the ontological and epistemological assumptions which underpin different research paradigms and strategies examined. Next, the methods utilised by researchers in construction management are examined through examination of a recent complete volume of the peer-reviewed journal Construction Management and Economics. This analysis reveals the extent to which methodological pluralism has been embraced by the research community to date. In addition, it examines the types of interpretative research methods applied by construction management researchers and questions. Thus, the results reveal both how the construction management research community has responded to the philosophical questions asked of it in the mid-1990s, and the diversity of research approaches that this has induced. The ensuing discussion speculates as to the likelihood of the insights gained through these research approaches informing the development and evolution of the industry that it serves. The likely impact of an enduring polarisation of philosophical position is juxtaposed against the potential benefits of multimethodological research design. This is used as the basis for the construction of an argument for the promotion of methodological pluralism in construction management research as a reaction to the entrenched views which seemingly pervade much of the community at present.

Research strategy and design

As a precursor to investigating the methods adopted by construction management researchers, it is necessary to review briefly the decisions which underlie research methodology, strategy and design. Clearly, research methodology in social enquiry refers to far more than the methods adopted in a particular study and encompasses the rationale and the philosophical assumptions that underlie a particular study. These, in turn, influence the actual research methods that are used to investigate a problem and to collect, analyse and interpret data. In other words, research methods cannot be viewed in isolation from the ontological and epistemological position adopted by the researcher.

In philosophy, ontology can be taken to broadly refer to conceptions of reality. Objectivist ontology sees social phenomena and their meanings as existing independently of social actions, whereas constructivist ontology infers that social phenomena are produced through social interaction and are therefore in a constant state of revision (Bryman and Bell, 2003: 19–20). Epistemology refers to what should be regarded as acceptable knowledge in a discipline (ibid. 13). Epistemological perspectives are bounded by the positivist view that the methods of the natural sciences should be applied to the study of social phenomena, and the alternative orthodoxy of interpretivism which sees a difference between the objects of natural science and people in that phenomena have different subjective meaning for the actors studied. Understanding the influence that competing paradigms have on the way in which research is carried out is fundamental to understanding the contribution that it makes to knowledge. Taking Bryman's (1988) definition of a paradigm as a 'cluster of beliefs and dictates which for scientists in a particular discipline influence what should be studied, [and] how research should be done', different research paradigms will inevitably result in the generation of different kinds of knowledge about the industry and its organisations. This perspective sees different paradigms as incommensurable, and so the choice of which paradigm to adopt fundamentally affects the ways in which data are collected and analysed and the nature of the knowledge produced.

In broad terms, the term 'research design' describes the ways which the data will be collected, analysed in order to answer the research questions posed and so provide a framework for undertaking the research (Bryman and Bell, 2003: 32). Making decisions about research design is fundamental to both the philosophy underpinning the research and the contributions that the research is likely to make. For example, qualitative research stresses 'ecological validity'; the applicability of social research findings to those that exist within the social situation studied. Choosing a reductionist approach to examining social phenomena (such as questionnaire survey) is likely to distance the enquiry from the social realities of the informant, thereby undermining its ecological validity. Thus, methods are inevitably intertwined with research strategy.

Without wishing to dichotomise or pigeonhole researchers within the construction management community, it is important to distinguish between the different types of research conducted as a backdrop to discussing the diversity of the methods employed. In broad terms, construction management research either adopts an objective 'engineering orientation', where the focus is on discovering something factual about the world it focuses on, or a subjectivist approach, where the objective is to understand how different realities are constituted (see Harty and Leiringer, 2007). Whilst the former emphasises causality and generalisability, the latter focuses on localised subjective meaning. In this chapter a distinction is also drawn between 'quantitative' and 'qualitative' research. Whilst this distinction is considered by some as unhelpful (see for example Layder, 1993; cited in Bryman and Bell, 2003), it nevertheless provides a useful framework for categorising the methods used by researchers. Indeed, it can be argued that quantitative and qualitative research are themselves rooted in particular ontological and epistemological foundations (i.e. objectivism and constructivism, and positivism and interpretivism respectively). Accepting this association between research methods and research paradigms enables philosophical differences in the role that theory plays in research to be viewed through the lens of the methods employed by researchers. In other words, the methods employed can be used as a proxy for the paradigm adopted. It is accepted that this represents an oversimplification of reality. For example, it is plausible that qualitative methods can be employed for theory testing as well as theory generation. However, as will be discussed later in this chapter, this is the case in the vast majority of construction management research projects.

The dominant research paradigm within construction management

In order to examine the methodological positions and research methods adopted by construction management researchers, an analysis was carried out of every paper published in *Construction Management and Economics* in Volume 24, 2006 (see Dainty, 2007). Each paper was scrutinised for statements as to the methodological position of the author(s) and the methods employed. Where this was not unambiguously stated within a defined section of the paper, efforts were made to identify the methods adopted from the narrative description of the research. In some cases, no discernable empirical research methods were adopted as the paper was a review-type contribution. In other cases, papers drew upon a multi-paradigm research design. These papers were defined as 'review' and/or 'mixed methods' respectively. Thus, four broad classifications were used for summarising the methodologies adopted within the papers as follows:

(1) *Quantitative* – unambiguously adopting quantitative methods rooted in a positivist research paradigm.
(2) *Qualitative* – unambiguously adopting qualitative methods rooted in an interpretative research paradigm.
(3) *Mixed methods* – comprising a combination of both inductive and deductive research methods.
(4) *Review* – not utilising empirical research methods.

For those papers which reported research which adopted a qualitative (2) or mixed method (3) approach, a further sub-classification step was undertaken to categorise the methods used. These categories were established inductively and were not based on an *a priori* classification of research methods. In this respect, the interpretation of the methods adopted by the papers studied is itself interpretative. This was necessary as some authors did not unambiguously state their adopted methods. The qualitative methods adopted by the authors comprised interviews (semi-structured and unstructured), focus groups and group interviews, observation (non-participatory and/or participatory including ethnography), document or other textual analysis and visual data analysis.

It is important to recognise several significant limitations of the approach adopted. Firstly, the papers published within *Construction Management and Economics* may not be reflective of the entire construction management research community. A search of papers published in other journals may have revealed that they attract papers from a different constituency of the research community which adopt different research methods. Secondly, this study represents an analysis of only those papers published and not *submitted* to the Journal. As such, the analysis may be more representative of the biases of referees rather than being necessarily representative of the methods actually adopted by construction management researchers. A third limitation concerns the nature of the methodological description contained within the papers themselves. This is highly variable and renders any such analysis somewhat tenuous. In addition, it is possible within some of the projects that other methods were employed which have not been unambiguously stated within the papers. These aspects may not have been published or may have been published elsewhere for legitimate reasons (such as word restrictions placed on articles within the Journal). A fourth issue concerns the reliability of drawing general conclusions based on a single year's worth of papers. It is possible that papers published in this year were anomalous to the general trends in the kind of papers published within the Journal. A final issue is that not all of the papers published within the Journal can be described as 'social research'. For example, some papers dealt with aspects of construction law or finance, which have only loose connections to social phenomena, for which the utilisation of qualitative methods would have been inappropriate. Despite these weaknesses, however, the Journal is considered by many construction management researchers to be one of the leading refereed publications in its field. This is supported by the very high levels of copy flow and the high rejection rate (see Taylor and Francis, 2007). Furthermore, it is reasonable to assume that, given that reviewers of papers are drawn from the construction management research community, that any bias towards methodological approaches would even itself out over time. The year selected for analysis, 2006 was the most recent year for which a full year's worth of papers were available. Furthermore, the Journal switched to a 12-issue format in 2006 which enables more papers to be considered in the analysis. Thus, whilst this chapter makes no claims as to the statistical reliability of the findings presented, and draws upon a wholly qualitative analysis of the narrative description of the methods employed within the papers, it does enable a simplified cross-sectional view of the dominant position of the research community.

The results of the analysis are presented in Tables 1.1 and 1.2. Table 1.1 presents an overview of the methods used within the research reported in the papers reviewed. These data represent the number of papers utilising the methods embodied by the broad classifications listed above. This shows that of 107 papers and notes published in Volume 24 of the Journal, 76 used quantitative methods. Only 9 used qualitative methods exclusively. In addition, a further 12 papers used a mixed methods approach combining qualitative and quantitative methods. It should be noted that in a few of the studies which have been classified as utilising exclusively quantitative approaches, a brief mention of exploratory interviews was made, although none of this was reported

	Qualitative methods	Quantitative methods	Mixed methods	Review/ other papers
No. of papers	9	76	12	10
(% within parentheses)	(8.4)	(71.0)	(11.2)	(9.4)

Table 1.1   Broad classification of research methods reported in all papers (excluding letters and book reviews) in Vol. 24 of *Construction Management and Economics* (n = 107).

	Interviews	Focus groups, workshops and group interviews	Observation	Document or textual analysis	Visual data
No. of papers	16	3	2	3	1

Table 1.2   Classification of research methods reported in papers using qualitative research methods in Vol. 24 of *Construction Management and Economics*.[1]

in the data. Although it could be argued that the qualitative findings may have shaped the resulting enquiry and quantitatively derived results, the fact that they did not warrant reporting in the papers provides justification for excluding them from the 'mixed methods' classification.

Table 1.2 presents a breakdown of the types of qualitative methods employed by those employing only qualitative methods and those adopting a mixed methods approach. In this table, papers have been classified under each category if the particular method has been utilised and the results reported in the paper. Thus, this table reflects the number of times that a method was applied across the sample of papers. Given that several studies employed a number of methods and datasets, this number is greater than the number of papers identified in Table 1.1. This table reveals that 16 of the 107 papers published in Volume 24 of the Journal used individual open-ended interviews. This represents more than three quarters of the studies employing qualitative methods. Three studies used focus groups, workshops and/or group interviews, two used forms of observation and three analysed documentary data (mainly as part of case study research). Only one paper reported analysing visual data.

Discussion: The implications of methodological uniformity

The construction management research community has clearly grown and developed since the methodological debates of the mid-1990s. This is reflected in the growth of the number of peer-reviewed journals and the numbers of papers published relating to the practice of construction management. Much of this work could be considered social science or sociological research, which is aimed at understanding the social structure and patterns of interaction between those working within, and affected by, the built environment and the agencies and institutions which structure it). Much of this work is also founded on the 'co-production' of knowledge. In other words, researchers use the real-world context of the industry as sites for developing research questions, and for conducting empirical work to examine them (Harty and Leiringer, 2007). It could be reasonably expected that their methodological positions and the methods adopted may have broadened and diversified to reflect the multiple traditions from which the community now draws. However, if the contents of this volume of *Construction Management and Economics* are reflective of the community at large, that is manifestly not the case. The findings raise fundamental questions, both in relation to the narrow ontological and epistemological standpoints of the research community, and in relation to the uniformity of methods that interpretive researchers employ.

Questions of social ontology are concerned with whether social entities are objective realities or social constructions built up from the actions and perspectives of social actors (Bryman and Bell, 2003: 19). It would seem on the basis of this analysis that the majority of construction management researchers have retained an objectified view of reality. Whilst it is by no means certain that the predominance of quantitative methods revealed in this paper is inexorably linked to positivist research philosophies (surprisingly, few of the papers actually stated a methodological position within the volume reviewed), it is highly likely that this reflects on on-going adherence to natural science methodologies and reductionist approaches to social enquiry within the community. Whether this should be seen as a concern will depend upon the individual standpoint of the reader, but the construction management community's apparent reluctance to embrace methodological pluralism has undoubted implications for the contribution it makes to both research scholarship and practice. It would seem that the research community has continued to adopt a rationalist paradigm in seeking to theorise on construction management as a discipline, with a resultant emphasis on causality over meaning (cf. Seymour and Rooke, 1995; Seymour *et al.*, 1997). Whilst it could be argued that the research community reflects, in microcosm, the industry's wider adherence to instrumentalist and rational solutions to complex managerial problems and situations (see Dainty *et al.*, 2007), it raises questions as to the ability of the construction management research community to be able to provide a rich and nuanced understanding of industry practice.

A second issue emerging from this analysis concerns the apparent reliance of qualitative construction management researchers on open-ended interviewing. As was discussed above, in contrast with quantitative research design, which remains relatively methodologically uni-dimensional, contemporary qualitative research is characterised by its diversity (Punch, 2005: 134). However, in the volume of *Construction Management and Economics* reviewed, virtually all of the studies which employed exclusively qualitative methods relied exclusively on semi-structured interviews. Within the social sciences, the apparent over-reliance on interviewing has been attracting criticism from researchers who see it both as symptomatic of the 'interview society' and as belying the fact that interviews are themselves methodologically constructed social products (see Hammersley and Gomm, 2005; cited in Gubrium and Holstein, 2002). In the past, those critical of interviewing have questioned their efficacy based on practical and pragmatic considerations such as the truthfulness of the informant and the differences between what people say and what they actually do (see Hammersley and Gomm, 2005). However, Hammersley and Gomm also point to a more radical critique of interviews as a research method as having recently emerged in which the social construction of what is said, and the fact that they reflect the particular context within which they take place, has been seen as limiting their methodological validity. Such a critique sees the interview informants as being more focused on self-presentation and the persuasion of others, rather than on presenting facts about themselves or the world in which they exist (ibid.). Regardless of whether such a radical perspective on the efficacy of interviews is fully accepted, the acknowledgement that they are in any way flawed reinforces the need for data from different sources to triangulate the inferences and outcomes that they provide.

An emergent finding also warranting further discussion concerns the lack of reflexivity within the papers reviewed. As was alluded to above, there is a tradition of reflexivity in qualitative enquiry where researchers openly question the effectiveness of their research methods on the robustness of their results and debate the influence and effect that their enquiry has had on the phenomena that they have sought to observe. Being reflexive requires a willingness to probe well beyond interpretation of the data, to explore how personal research bias affects the research process itself (Woolgar, 1988 ; cited in Bryman and Bell, 2003: 529). Despite this however, there is an absence of critical reflection in many of the papers reviewed which adopted qualitative methods. This may reflect that dominant rationalist paradigm of the construction management research community, or even a concern on the part of interpretative researchers that such a reflection would effectively equate to an admission of 'flaws' in their research designs. However, an apparent reluctance to engage in reflexivity arguably has a detrimental effect on the methodological evolution of the discipline and the development of its theoretical base.

The case for methodological pluralism in construction management research

In charting the history of pluralism, Mingers (1997: 3) notes that philosophers such as Hanson, Kuhn and Popper demonstrated flaws in the cornerstones of induction, and theory- and observer-independent observation. He argues that in social science, this legitimated the emergence of the various schools of interpretivism such as phenomenology and hermeneutics. He also notes that similar trends emerged in management science in the 1980s with the emergence of soft systems methodology (SSM) and other soft operations research (OR) approaches. It was through the challenge to the positivist orthodoxy by the emergence of phenomenological and structuralist epistemological positions that the new perspective of 'methodological pluralism' emerged.

The basic principle of methodological pluralism is that the use of multiple theoretical models and multiple methodological approaches is both legitimate and desirable if established models and understandings are to be questioned and knowledge furthered. Adopting the principles of methodological pluralism does not render the choice of method arbitrary, but emphasises the context-sensitivity inherent in research design. Indeed, many researchers argue that quantitative methods should be combined because theory building required 'hard' data for uncovering relationships and 'soft' data for explaining them (see Loosemore et al., 1996).

Mingers' (1997: 9) methodological pluralism may be considered in three ways. Loose pluralism suggests that a discipline should support and encourage a variety of paradigms and a range of methods without prescribing how they should be used and applied. Complementarism views regarding different paradigms as internally consistent and therefore more or less appropriate for a particular situation. Strong pluralism holds that most situations are best dealt with by a blend of methodologies originating from different paradigms. In a similar vein, Hammersley (1996; cited in Bryman and Bell, 2003: 482) classifies multi-strategy research into three broad approaches. 'Triangulation' refers to the use of qualitative research to corroborate quantitative research (or vice versa); 'Facilitation' is where one research strategy is employed in order to aid research using another approach; and 'Complementarity' is where two strategies are employed in order to dovetail different aspects of an investigation. In management science research, Complementarism (cf. Flood and Jackson, 1991) concerns the selection of a methodology for a particular intervention rather than the combination of parts of methodologies together (Mingers and Gill, 1997: xv). The practice paradigm of linking of different *aspects* of methodologies has been termed 'multimethodology' by Mingers and Gill, and in many respects exemplifies the principles of methodological pluralism. Indeed, Mingers (1997) refers to this principle as 'strong pluralism' because of its emphasis on blending methodologies from different paradigms within a single intervention.

The theoretical attractiveness of multimethodology lies in its abilities to enable the handling of problematic situations which require the effective linking of judgement and analysis (Rosenhead, 1997). In other words, it provides a framework for utilising the plurality of methodologies in order to understand or intervene in a complex situation. Given the inherent complexity of the construction industry as an arena within which to conduct research, and the problem-focused orientation of construction management research (see Harty and Leiringer, 2007), the theoretical benefits of multimethodology seem obvious. Thus, in some respects the future development of construction management research will depend upon the willingness of its research community to see qualitative and quantitative research as complementary rather than competitive and mutually exclusive (Loosemore et al., 1996).

The analysis presented above also reveals the narrowness of the methods employed in construction management research. A shift towards multimethodological perspectives on research design brings with it a need to embrace a greater multiplicity of different methods. For construction management researchers this will mean a greater emphasis on qualitative enquiry. There is no room within this chapter for an in-depth treatise on the multiplicity of methods that fall under the broad heading of qualitative research (see Denzin and Lincoln, 2000; Cassell and Symon, 2004; and Silverman, 2004, 2005 for this). Moreover, it is important to stress that qualitative research is a complex, changing and contested field (Denzin and Lincoln, 2000) which is characterised by its diversity (Punch, 2005: 134). Indeed, some writers have criticised attempts to classify qualitative research as a generic approach to doing research (Silverman, 1993). Nonetheless, a broader outlook with regards to the application of research methods is a prerequisite for embracing the principles espoused above.

Challenges in undertaking multi-paradigm research

As could be expected given the polarised debate which divides those in the positivist and interpretivist camps, combining methodologies is not without its critics. Indeed, although this chapter has advocated multi-paradigm and multi-strategy research approaches, combining methods and methodologies is by no means a straightforward undertaking. A range of philosophical, cultural and psychological hurdles confront the multi-paradigm researcher, each of which renders it a highly problematic undertaking.

According to Bryman and Bell (2003: 480) the argument against multi-strategy research methods essentially rests on two arguments. Firstly, research methods carry epistemological commitments. The embedded nature of methods is such that they are inexorably connected to the views of the world from the paradigm from which they originate. This 'paradigm incommensurability thesis' suggests that researchers must choose the rules under which they undertake research based on the fundamental assumptions that they bring to their enquiry (Mingers, 1997: 13). Thus, seeking to understand a practitioner's perspective on a situation is consistent with interpretivism, but inimical to positivism. A second challenge is that quantitative and qualitative research represent *separate* paradigms. In other words, quantitative and qualitative approaches are underpinned by different assumptions and methods which are incompatible between paradigms. Of course, these arguments are predicated on the view that quantitative and qualitative research are in fact research paradigms, even though areas of commonality exist between them. Nevertheless, as was discussed above, research methods tend to be rooted in particular epistemological positions.

Mingers (1997: 14–15) takes this line of thinking further in problematising the linking together of research methods across different research paradigms. He suggests that paradigm sub-cultures exist within management science disciplines which are shaped by the backgrounds of researchers. Individuals' methodological preferences will be reinforced by the institutional, physical and geographic boundaries around which they coalesce. Mingers also points towards cognitive barriers in that predilections towards particular paradigms may be so entrenched as to prevent the adoption of seemingly competing philosophical standpoints. Given this backdrop, it is little wonder that most researchers nail their colours to a particular philosophical mast and root their work within a distinct methodological paradigm. The danger for those eschewing the tendency to position themselves in a particular camp is to run the risk of finding themselves in a methodological 'no mans land'! Thus, those embarking on this journey must have the courage to challenge the historical values which have hitherto maintained the paradigmatic intransigence of those on both sides of the epistemological divide. But it is only by demonstrating the potential of methodological pluralism that entrenched attitudes are likely to shift, and a richer understanding of the practice of construction management and the workings of the industry's organisations and projects is likely to emerge.

Conclusions

This chapter has discussed the implications of the apparent narrowness of the construction management research community's methodological outlook and the implications for understanding of the practice of construction. The construction management field appears to be firmly rooted within the positivist tradition. It has shown both an entrenched adherence to positivist methods within the community, and a significant reliance on open-ended interviews by those adopting qualitative methods. Clearly, no claims can be made as to the broader significance of these findings as they do not provide evidence of methodological trends. However, given the methodological debates of the mid-1990s, they do provide limited evidence of an apparent reluctance to embrace paradigmatic change. Moreover, they present a view of a community reluctant to adopt the kinds of radical qualitative research methods which could provide richer insights into industry practice. The apparent lack of methodological diversity, coupled to an apparent lack of adventure in interpretative research design, suggests a research community rooted in methodological conservatism and disconnected from the debates going on in many of the fields from which it draws. An enduring adherence to the positivist paradigm will do little to enable construction management researchers to grasp the meaning of social action from the perspective of the actors involved.

Many of the research approaches, methodologies and methods espoused within the other chapters of this book offer routes for addressing the problems alluded to within this chapter. It has been argued that those engaged in social science research in construction management could usefully embrace multi-strategy or 'multimethodology' research design in order to better understand the complex network of relationships which shape industry practice. This radical perspective eschews traditional dualisms by suggesting that no single methodology can ever provide a complete picture of the projects and organisations that form the arenas for construction management research. Adopting a diversity of approaches would move the construction management research community towards a more balanced methodological outlook and would begin to challenge the dominant positivist paradigm which seems so pervasive within the community. This is not to suggest that there is no place for positivism in construction management research, but that used in isolation such perspectives do not provide the types of insights required. As Mingers (1997: 9) states

> Adopting a particular paradigm is like viewing the world through a particular instrument such as a telescope, an X-ray machine, or an electron microscope. Each reveals certain aspects but is completely blind to others ... each instrument produces a totally different, and seemingly incompatible, representation. Thus, in adopting only one paradigm one is inevitably gaining only a limited view of a particular intervention or research situation ... it is always wise to utilize a variety of approaches.

Advocating the combination of methodologies rejects some of the traditional dualisms which have seemingly pervaded the discourse of how we should undertake construction management research in the past ten to fifteen years. As has been explained however, the benefits of holism – combining methodological perspectives in order to gain richer insights and a more complete understanding of social phenomena – are particularly persuasive in the context of doing research in the construction sector. A more expansive outlook towards mixing methodologies and research paradigms could yield deeper insights into, and understanding of, the way that practitioners 'do' management in the construction sector. Techniques such as triangulation, facilitation and complementarity (cf. Hammersley, 1996) all offer the potential to overcome the weaknesses of single-paradigm approaches, whilst multimethodology – the combination of parts of methodologies together – offers particular advantages for the use of systems or operational research techniques (Mingers and Gill, 1997). However, mixing paradigms in this way will require adventure and courage on the part of researchers if they are to challenge the paradigmatic intransigence which is seemingly so ingrained within the construction management research community.

Acknowledgements

An earlier version of this chapter, including the empirical aspects, first appeared in the *Construction Management and Economics* 25th anniversary conference, University of Reading, 2007 (see Dainty, 2007). An abridged version of this chapter was also presented as the keynote presentation at the Postgraduate Researchers of the Built & Natural Environment (PRoBE) conference, Glasgow, 2007.

Note

1 It should be noted that on several occasions the nature of the data collected and the methods employed was ambiguous. Although a reasoned judgement has been made based on the contents of the paper and the nature of the results, the accuracy of these assertions cannot be guaranteed given the lack of detail within the papers.

References

Betts, M. and Lansley, P. (1993) Construction management and economics: A review of the first ten years, *Construction Management and Economics*, 11(4), 221–245.

Bryman, A. (1988) *Quantity and Quality in Social Research*, Routledge, London.

Bryman, A. and Bell, E. (2003) *Business Research Methods*, Oxford University Press, Oxford.

Cassell, C. and Symon, G. (2004) *Essential Guide to Qualitative Methods in Organizational Research*, Sage, London.

Dainty, A.R.J. (2007) A review and critique of construction management research methods, in Hughes, W. (ed.) *Proceedings of Construction Management and Economics 25th Anniversary Conference*, University of Reading, 16–18 July, p. 143.

Dainty, A.R.J., Green, S.D. and Bagilhole, B.M. (2007) People and culture in construction: Contexts and challenges, in Dainty, A., Green, S. and Bagilhole, B. (eds) *People and Culture in Construction: A Reader*, Taylor & Francis, Oxon, pp. 3–25.

Denzin, N.K. and Lincoln, Y.S. (eds) (2000) *Handbook of Qualitative Research* (2nd Edition), Sage, Thousand Oaks, CA.

Fellows, R. and Liu, A. (1997) *Research Methods for Construction*, Blackwell, Oxford.

Flood, R.L. and Jackson, M.C. (1991) *Creative Problem Solving*, Wiley, Chichester.

Gubrium, J.F. and Holstein, J.A. (2002) *Handbook of Interview Research*, Sage, Thousand Oaks, CA.

Hammersley, M. (1996) The relationship between qualitative and quantitative research: Paradigm loyalty versus methodological electism, in Richardson, J.T.E. (ed.) *Handbook of Research Methods for Psychology and the Social Sciences*, BPS Books, Leicester.

Hammersley, M. and Gomm, R. (2005) Recent radical criticism of the interview in qualitative inquiry, in Holborn, M. and Haralambos, M. (eds) *Developments in Sociology*, Vol. 20, Ormskirk, Causeway Press.

Harty, C. and Leiringer, R. (2007) Social science research and construction: Balancing rigour and relevance, in Hughes, W. (ed.) *Proceedings of Construction Management and Economics 25th Anniversary Conference*, University of Reading, 16–18 July.

Harriss, C. (1998) Why research without theory is not research. A reply to Seymour, Crook and Rooke, *Construction Management and Economics*, 16, 113–116.

Layder, D. (1993) *New Strategies in Social Research*, Polity, Cambridge.

Loosemore, M., Hall, C. and Dainty, A.R.J. (1996) Excitement, innovation and courage in construction management research – Challenging historical values, in Thorpe, A. (ed.) *Proceedings of the 12th Annual ARCOM Conference*, Sheffield Hallam, pp. 418–427, ISBN: 0 86339 660 7.

Mingers, J. (1997) Multi-paradigm methodology, in Mingers, J. and Gill, A. (eds) *Multimethodology: The Theory and Practice of Combining Management Science Methodologies*, Wiley, Chichester, pp. 1–20.

Mingers, J. and Gill, A. (1997) *Multimethodology: The Theory and Practice of Combining Management Science Methodologies*, Wiley, Chichester.

Punch, K. (2005) *Introduction to Social Research* (2nd Edition), Sage, London.

Raftery, J., McGeorge, D. and Walters, M. (1997) Breaking up methodological monopolies: A multi paradigm approach to construction management research, *Construction Management and Economics*, 15(3), 291–297.

Rooke, J., Seymour, D. and Crook, D. (1997) Preserving methodological consistency: A reply to Raftery, McGeorge and Walters, *Construction Management and Economics*, 15, 491–494.

Root, D., Fellows, R. and Hancock, M. (1997) Quantitative versus qualitative or positivism and interactionism – a reflection of ideology in the current methodological debate, *Journal of Construction Procurement*, 3(2), 34–44.

Rosenhead, J. (1997) Forward, in Mingers, J. and Gill, A. (eds) *Multimethodology: The Theory and Practice of Combining Management Science Methodologies*, Wiley, Chichester.

Runeson, G. (1997) The role of theory in construction management research: Comment, *Construction Management and Economics*, 15(1), 299–302.

Seymour, D. and Rooke, J. (1995) The culture of the industry and the culture of research, *Construction Management and Economics*, 13(6), 511–523.

Seymour, D., Crook, J. and Rooke, J. (1998) The role of theory in construction management: Reply to Runeson, *Construction Management and Economics*, 16, 109–112.

Seymour, D., Rooke, J. and Crook, J. (1997) The role of theory in construction management: A call for debate, *Construction Management and Economics*, 15(1), 117–119.

Silverman, D. (1993) *Interpreting Qualitative Data: Methods for Analysing Qualitative Data*, Sage, London.

Silverman, D. (ed.) (2004) *Qualitative Research: Theory, Method and Practice* (2nd Edition), Sage, London.

Silverman, D. (2005) *Doing Qualitative Research: A Practical Handbook* (2nd Edition), Sage, London.

Taylor and Francis (2007) *Statistical Summary of Annual Performance for Construction Management and Economics*, available at http://www.tandf.co.uk/Journals/pdf/announcements/rcme_stats.pdf.

Woolgar, S. (1988) *Science: The Very Idea*, Ellis, Chichester.

Chapter Two

Architectural research

Alan Penn

Introduction

Architectural research forms an integral part of the field of built environment research, representing as it does one of the key stages in the production and evolution of the built environment. It is increasingly hard, in the modern world, to conceive of innovation in the built environment without consideration of the architect, architectural design processes and the buildings and urban areas that these produce. In fact, it seems impossible to account for certain of the great failures of recent built environment history without invoking architectural theory and architectural practice in some way. Equally, one could argue the impossibility of accounting for some of the greatest innovations and successes without giving credence to the role of architectural design as an objective factor. The aim of this chapter, however, is broader than just to describe architectural research as an academic activity aimed at throwing light on an undoubtedly important component in the production and management of the built environment; the aim is to show how architectural research in all its diversity is integral to architectural practice itself, as well as to the powerful role which, it will be argued, architecture plays in shaping the evolutionary pathways taken by communities, cultures and organisations.

This chapter starts by sketching architectural design as a subject area and as a professional practice in relation to the production and operation of the built environment. It uses this sketch to propose those features of knowledge which might be considered specifically architectural in nature, and then goes on to describe how these have given rise to a landscape of architectural research approaches and methodologies. At this point, distinctions are drawn between substantively different approaches rather than trying to cover the variations in terminology held for essentially similar programmes of research.

The third section deals in greater detail with a specific field of architectural research – the analysis of spatial layout in plan – arguably the main product of architectural design in so far as buildings play an instrumental role aside from that of providing shelter from the elements – and the investigation of its relation to aspects of the social and economic performance of the built environment. The selection of this specific field is far from arbitrary. It provides a unifying framework within which other aspects of architectural, planning and engineering research can be synthesised, and it defines a natural linkage to the field of design practice and innovation. A set of research methods for investigating spatial layout in plan is outlined and a brief overview is given of the main substantive findings of this area of research. In this way it is argued that architectural research forms a consistent if complex field, in which historical studies and critical theory serve to translate into language, and so open for discussion, the primarily non-verbal projects and buildings – the products of practice – which form the focus for architectural innovation. The relation between innovation in the social, economic and cultural forms produced by the built environment and the technological, material, structural and environmental means through which these forms are realised is shown to be part and parcel of the same process. Here architectural research is shown to be not merely a passive 'observer' or 'explainer' or architectural phenomena, but to be actively embedded into the whole culture of architecture and the way that this works to foster social reproduction and innovation.

A sketch of architectural design

It is a general characteristic of architectural design that the brief is poorly defined at the outset. It is this that leads it to differ as a process from many other fields of design which are directed towards finding a solution to a relatively well-defined problem. In architectural design, the process is one of co-evolution of the design and the brief with the one stimulating the other iteratively: formal possibilities are explored through design options, the implications of these as solutions for possible brief requirements are used to elicit better framed requirements, statements from clients and users; these in turn stimulate refinements or new design options. At the end of the design process, one should have developed both a design and a relatively well-stated brief.

This assertion may seem like heresy to those versed in the primacy of the requirements, statement, but there are two very good reasons why the architectural brief is ill defined. First, architecture operates in a highly complex and holistic problem space. A single built form must function within a broad set of independent domains: structure, shelter, servicing, aesthetic, social and economic. It is a product composed of a wide range of technologies and systems, and will be procured within a complex environment of craftsmen, firms, markets and regulations. Although the functional requirements are relatively independent – for example, the laws of physics according to which structures operate and those of a building's socio-economic performance are entirely different – these systems interact through the building itself. The single building must be designed to achieve functional ends for all of these different domains. If in order to achieve the objectives of one domain we change the building design, that change will have implications for all other domains. The sheer complexity of the built product, the independent but interactive nature of this function, and the complex processes of its production make it hard to write a formal specification at the outset for all but the simplest of building types.

This complexity is exacerbated (from the point of view of brief writing) by the precise way in which the single design solution must interact with all these domains of function. For a given design solution, certain aspects of a building's performance may turn out to be 'not much of a problem'. Indeed, this can be seen as a fundamental aim of modernist architectural design. When Ludwig Mies Van de Rohe said 'less is more', he was alluding to this property of architectural design solutions – to eliminate or answer several sets of problem with a single design move. However, what this means is that the specific brief for a building is in effect a function of the chosen solution strategy. This makes the relationship between briefing and design more than merely iterative, but in a strong sense complex and emergent. The feedback between a given design solution and a brief statement, as these iterate through the design process, leads to bifurcations in the developmental pathway of both the evolving design and the evolving design brief. At particular points in that evolution, a strategic event may occur which throws the evolutionary trajectory onto a new pathway. These events may be either design decisions (selection of a specific option from those available at that point in time), or briefing discoveries (such as the identification of important requirements or constraints, or the discovery that a particular design approach renders certain aspects 'not a problem', allowing them to be removed from the brief entirely), but they share in common a strong determining effect on the future design pathway.

Second, and perhaps more fundamental, is the fact that many aspects of a building's function are 'non-discursive' (Hillier, 1996). That is, buildings construct environments that we experience largely subconsciously and which often we do not have language to describe. Certainly for the non-professional it can be more or less impossible to describe in language what it is that we want to achieve in a building, even if we consciously understand that. More often our understanding is itself vague and apparently intangible. A retailer may speak of the importance of the 'customer experience', a research scientist of an 'innovative atmosphere'; both critically important outcomes of a successful building project, but both apparently indefinable except as judgements of a building in retrospect. It is for this reason that expecting a client or user to write an explicit and well-defined brief at the outset is not only unreasonable, but is often counterproductive. It is also for this reason that much of the body of knowledge about architectural design is tacit and transferred through a community of practice, rather than didactically taught and explicitly set out. This defines the nature of architectural design education, and explains why it differs so markedly from other disciplines.

The tacit nature of architectural practice and its need to deal with the intangible aspects of building performance also helps to explain why it is a profession. The professional can be defined as an individual who takes personal responsibility for another's decisions in areas of high uncertainty and risk (Jamous and Pelouille, 1970). If knowledge were predominantly codified and explicit, a more rule governed and bureaucratic structure would be natural, however, in a field as complex as architecture, one has to rely on the professional's trained intuition and their experience of previous cases to deal with the vitally important but intangible aspects of building performance.

The organisational structures of architectural practice have evolved to deploy both tacit and scientific knowledge in order to manage complexity and uncertainty. Amongst these structures are strong divisions of labour in the domain, formed as knowledge moves from tacit to explicit. For example, as knowledge of structures became scientific and principled in the nineteenth century it became vested in the structural engineer; the same can be said of building physics and environment, construction and management, or procurement. A common mode of operation today is through a 'project team' composed of domain experts, various kinds of engineer and surveyor, each with well-defined and more or less explicit fields of knowledge. The architect's position around this table, however, is somewhat different. His role is to propose a design solution which other domain experts then comment upon from their own field of expertise. The reason for this is that the architect remains responsible for finding a single design solution that works for (or perhaps renders unproblematic) the whole range of different and substantially independent problem domains. The design process is characterised by iteration through options in which the field of design possibility (covering different solution strategies) is explored and commented upon from various viewpoints. This is a collaborative process of exploration of a multi-criteria problem space in which both the problem statement (the brief) and the solution strategy (the design) evolve together, but in which both explicit and tacit bodies of knowledge held by different individuals are applied.

The architect also tends to retain responsibility for the more intangible factors – the social, cultural and aesthetic for example – that remain the province of application of tacit knowledge in the process. This area of practice is worth examining and, in particular, it is worth considering the weight given in contemporary culture to the 'aesthetic' or 'style' of the individual architect or their firm. The pursuit of an individual signature, a visual or stylistic aesthetic, or an ethics or value-based position or set of processes, is perhaps the greatest driving force in contemporary architectural culture. At the same time culture seems, sometimes, to evolve as a fashion industry. One year the fashion may be for symbolic and iconic forms, the next it is for expressive and organic structures, or a 'green' ethos. Here we have innovation and change which appears to drive the individual creative direction of architectural culture. It is also an area which dominates much discussion within individual practices. There are two reasons for this. First, questions of style, approach and ethos underpin a practice's signature and so become a differentiator in the marketplace. Second, and more important, these factors add an important layer of constraint on the infinite field of possibility faced by the designer. Matters of style and approach set a starting point for the design process and act as a significant factor in shaping the path a design will take. This becomes apparent in the built works of a practice as well as their unbuilt projects, and allows a client to select, not a specific solution since every building is unique, but an architect whose processes, style and approach seem to give the kind of results which that client, intuitively, feels will be right.

As a practice grows beyond the size in which the founding partners can control the design approach, quality and style of every building, mechanisms must be developed to allow that control to be established. Bill Hillier has famously described one model in terms of a 'propose-dispose' or 'Hamadryas baboon' social structure.[1] In this, the senior partners move to a position of agreeing or rejecting propositions from aspiring junior staff. This can be seen as a version of a community of practice, in which implicit stylistic rules are promulgated tacitly and evolve through proposal and disposal, rather than explicitly through well-defined rule systems.

The role of the aesthetic as well as the ethical position taken by a practice can be seen to be highly influential. A practice may take a position on, say, environmental sustainability which will define much of its design approach and work processes, in the same way that they might adopt the Miesian 'less is more' aesthetic. These positions ultimately define the work of a practice, and are the subject of selection by clients. However, they are also reflections of a wider architectural culture, and it is the role of architectural critics and historians to identify and give voice to these wider trends in contemporary culture. Critical discourse is one of the most influential mechanisms through which architectural theory and research informs the world of practice.

The structure of architectural research

Two broad divisions must be made in contemporary architectural research. First, between those aspects of research that deal primarily with process and those that deal with product; and second, between those that deal with explicit or scientific knowledge and those that deal with tacit bodies of knowledge. The long history of progress in research into the architectural product has been to move from tacit knowledge to increasingly explicit knowledge. Fields of engineering, building science and materials science have become increasingly well understood formally, and are rigorously underpinned by explicit scientific theory, as, too, are those in process and management. However, behind all these lie the multiple socio-economic, cultural and aesthetic 'intangible' aspects of building performance. In these aspects of architecture progress is subject to innovation through practice. It is here that drawn, but unbuilt, projects can be as influential in the long term as buildings themselves. The Archigram projects of the 1960s and the 1970s still have an impact on practice today, as do the unbuilt works of Mies and Corbusier. The speed with which novel architectural forms propagate in the contemporary scene shows how influential innovation of this kind can be.

In the area of architectural innovation through practice the non-discursive nature of the building exerts a considerable effect. It is primarily because of this that the architectural historian and the critic play such a key role in architectural culture. Their role is to translate buildings and projects from the non-discursive world of experience into the world of language, and so to bring them within a discursive culture. The historian and critical theorist draw together buildings and projects designed in isolation; by investigating parallels they define trends and contradistinctions, and reveal the generic principles that can be derived from a heterogeneous field of work. In doing so they help to create architectural culture as something that is shared between practitioners. The impact of contemporary critical theories and discourse on design practice should not be underestimated. This often provides the grit which stimulates designers along particular pathways, helping to give voice to the ethical and aesthetic positions which inform their practice. However, it is the buildings and projects themselves which ultimately form the material for both critical commentary and which, through published drawings and photographs, directly influence contemporary architectural culture. This is noticeable in terms of fashionable forms and details, and the way that these change over time, but also in the way that critical texts and professional journals define the contemporary fashion and design milieu.

The field of critical discourse forms a part of the milieu within which contemporary design takes place. However, this is a particularly risky business. The speed with which new ideas propagate can often itself create problems if those ideas are in any sense poorly founded. The experience of post-war social housing and urban planning demonstrates this clearly. Think for example of the notion of 'territoriality' and its interpretation in terms of 'defensible space', or the 'gravity theory' of urban function and its legacy in urban traffic engineering and land use planning. Wrong-headed ideas such as these have led to a legacy of social and economic dysfunction that still blights our cities. The next section describes a field of architectural research that emerged in the late 1970s in the UK in direct response to this history.

Space syntax and the social logic of space

During the late 1960s and early 1970s it was beginning to be recognised that the programme of functionalist modernism had given rise to some of the greatest functional failures of architecture and urbanism that had yet been seen. These failures were not in the main failures of structure or fabric, still less of construction process and management. If anything, it was quite the reverse. Innovations in construction processes allowed one of the most efficient periods of building to take place in the post-war construction of social housing. Of course failures of structure and fabric did occur as a consequence of technological innovation, however these were quickly diagnosed and regulated against since, by and large, they were open to scientific explanation. The systematic failures were more chronic and social in their form. Large-scale housing schemes gave rise to 'new town blues' alongside a suite of problems of social malaise. Mono-functional urban areas led to descriptions of the 'urban desert' and 'perpetual night'. The reaction in architectural circles was most often to blame housing management for loading estates with problem families and for failing to properly provide the social, economic and educational support that these families needed. The reaction from those that lived on the estates was different. For them it was clear that the architecture was to blame. It was against this background that Bill Hillier and his colleagues at the Bartlett, UCL, set up their programme of research (Hillier et al., 1983).

It seemed clear that if architecture were to blame for the social failures that were being experienced, a new kind of theory must be developed in which society and the design of the built environment could be linked together. It seemed equally clear that what was needed was not a normative theory of how space or society *ought* to be, since there was no lack of such theories and indeed the failing designs seemed to be based on one or another, but an empirically based and analytic theory that would help explain the successes and failures of design (Hillier *et al.*, 1972).

Finding a basis for an empirical theory in architecture poses a serious problem: which aspects of architecture should one choose as in any sense determinative of social function and failure? Hillier's answer was simple: spatial configuration (Hillier and Hanson, 1984). This is the pattern of spaces and relations between spaces which compose buildings and cities. It is space through which people move, and in which they are brought into face-to-face contact with one another. Contact in turn is a prerequisite for interaction, communication and the transactions of social and economic life. In this way it seemed possible to make the link between design and social outcomes.

The proposition was that the geometry and network topology of spatial patterns formed by the built environment directly affects patterns of movement, and so co-presence and interactions between people. It therefore has direct social and economic effects and these should be open to investigation through careful observation and analysis. The programme of research developed methods for representing and quantifying the geometric and topological properties of space patterns in order to allow differently planned buildings and urban areas to be compared on a quantitative basis. This allowed 'design' at the level at which it affects patterns of space through which we move and in which we meet or are kept apart to become subject to quantitative analysis and comparison.

The methods of analysis developed for this research pass through a number of stages. First, because space in the built environment is continuously accessible it was decided that one should decompose it into a series of somehow equivalent 'spaces' which could then be analysed in terms of how they relate one to another. Inside buildings this process can seem fairly straightforward. A space is a 'room' and a doorway or threshold links two such spaces. This allows the construction of a simplified network of 'nodes' and links that represent the basic patterns of accessibility and circulation within a building plan. Since social functions often map onto relatively well-defined spaces or rooms in buildings, analysis of room relationships, represented as the graph of nodes and links, can allow one to investigate the way social relations are constructed in space. An important aspect of this is that different spaces in a single building are differently embedded in the network, and this is measurable. In Figure 2.1 the 'salle commune' or main eating space of a French vernacular farm house is found also to be spatially shallow whilst the formal 'grande salle' is more isolated. This is not chance, but has been shown to be systematic in domestic space layout for this class of buildings in this part of France (Hillier *et al.*, 1987).

In thinking about this it is worth considering the different options open to an architect to design a building layout in different ways – if you like, what possibilities does a building plan designer's palate offer? Some aspects of design are very obvious – spaces can be more or less well defined and they can be small or large, however the network properties of a building plan are more complex. To simplify the account greatly, a network of spaces can be characterised firstly by the degree to which spaces connect together to form rings or circuits in the circulation system, or conversely the degree to which that system is treelike and one is forced to retrace one's steps to get from part to part; and secondly by the amount of depth – the number of intervening spaces – in

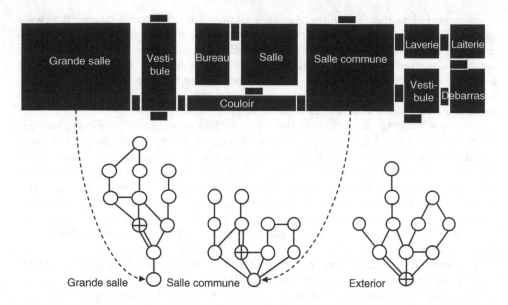

Grande salle · Salle commune · Exterior

Figure 2.1 A simple house plan, its representations as rooms and as a network (Hillier *et al.*, 1987). The same network can be arranged in steps of depth considered from the point of view of different spaces, showing how a single system of spaces is objectively different from different points of view.

the network. Networks can be relatively shallow where spaces are directly connected or there are only a few intervening spaces, or they can be extremely deep (Figure 2.2). Both ringiness and depth can also be thought about in terms of whether they are constructed locally or globally in the network, and different platforms can vary in the way that each of these properties are realised.

There is at least one other important aspect of building plan layout on the architect's palate. This is the geometric property of 'lining spaces up' or creating chicanes. It is possible for example to take a series of well-defined rooms and then align their doorways 'enfilade' so that a single line of sight or circulation passes directly through all of them. Alternatively, doorways can be offset so that one's view is never more than from one space to the next (Figure 2.3). This is dealt with analytically by representing these lines of sight and access in the form of a second map – the axial map (Figure 2.4). All the longest lines of sight and access are drawn that pass through rooms and thresholds, and which connect to make any rings of circulation. These are again represented as a network in which lines are considered as 'nodes' which are linked to other lines they intersect, allowing their properties as graphs to be measured. In large buildings such as hospitals or schools, as well as in urban space, this representation allows the key role of linear corridor systems and streets to be represented and analysed.

All of the morphological possibilities defined by the 'architect's palate' can be exploited for social purposes and the architect's job can be seen as developing a building layout which will be appropriated for meaningful use by the building users according to the affordances offered by space, geometry and spatial relations. Architectural research using these techniques can then be seen as aiming to uncover any systematic ways in which building users appropriate space for social use, and how this becomes meaningful in culturally specific ways. Since architecture is non-discursive, and both lay

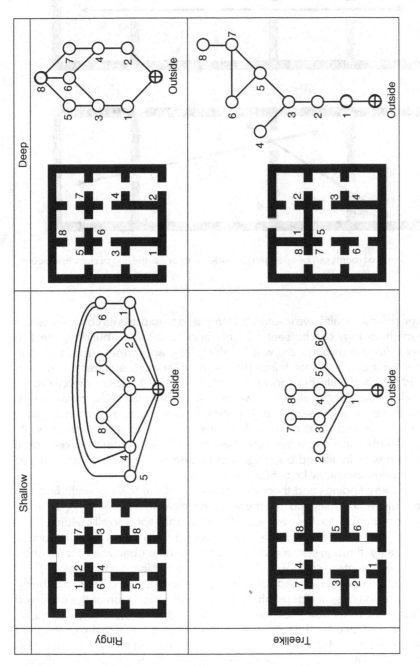

Figure 2.2 Four plans and their networks showing variations in ringiness and depth (Hanson, 1999).

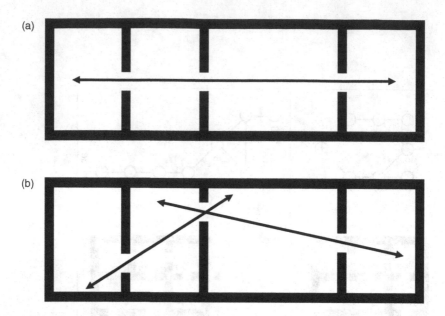

Figure 2.3 A sequence of rooms can be arranged 'enfilade' or to break the line of sight from room to room.

users and design professionals have insufficient language to express its complex nature in words, the methodology can be seen as a method for allowing building users to voice their views directly through the way in which they appropriate space in their everyday behaviour. Equally, research into the *oeuvre* of specific architects can interrogate the buildings they built (or proposed), and allow these to speak directly to us, often giving the lie to what those architects wrote or said about the selfsame buildings. Comparative and historical research can investigate differences and changes over time in the same way, and on the basis of a consistent set of representations. At one extreme, archaeological researchers use these techniques to infer the likely social structures of those who inhabited buildings and settlements in which little remains other than information on spatial layout (Bustard, 1999).

There are a four key findings and theories that emerge from this research: first, the theory of 'natural movement'; second, the theory of the 'movement economy'; third, the concept of 'generic function' as those aspects of function that apply to all buildings; and lastly, the notion that one of the main social outcomes of architecture is the construction of a 'virtual community' through systematically structuring the probabilities of encounter between different segments of society or different groups within organisations.

Space syntax researchers have found consistent evidence that, all other things being equal, the configuration of space determines a significant proportion of the observed variation in movement flows from location to location in spatial systems (Figure 2.5). Flows are also affected by other factors, such as the location and density of different space/land uses or attractor facilities. However 'natural movement' is defined as that proportion of movement in a system that can be explained in terms of the configuration of the spatial system alone, without invoking the effects of attractors or generators of movement, or rule systems operating on human behaviour (the school timetable for example). Natural movement forms the background movement pattern that might be expected to result from a specific plan form (Hillier *et al.*, 1993).

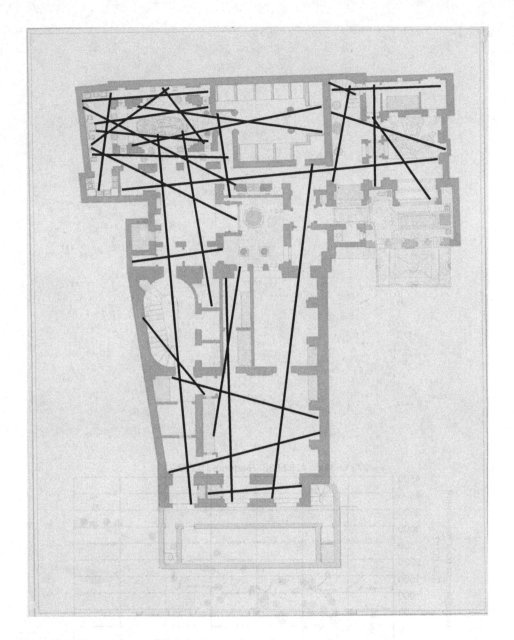

Figure 2.4 The axial map of Sir John Soane's house basement plan.

An immediate consequence of natural movement is that certain spaces, purely as a result of their location with relation to all other spaces in the system, gain through-movement. This has a direct economic consequence, particularly when we consider urban space (but also within retail buildings), that these spaces generate a higher footfall, and so are more valuable for retail land uses. This in turn stimulates agglomerations of retail land uses which then become the destination for new trips, generating more movement in other spaces in the system. This type of process is strongly emergent in that it involves feedback between location, land use, development

(a)

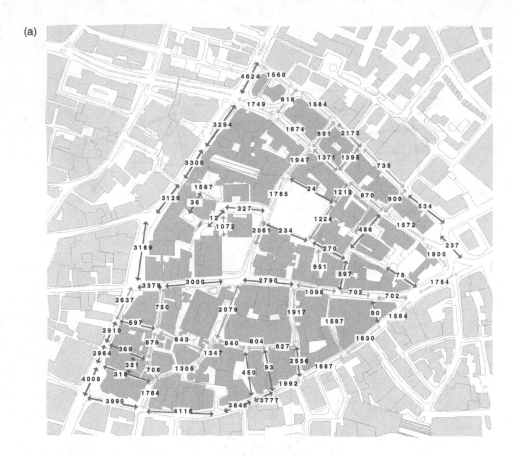

(b)

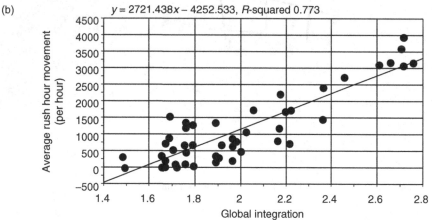

Figure 2.5 (a) Baltic House area of the City of London marked with rush hour average flows (Adults/ph); (b) Correlation between global integration and rush hour pedestrian movement, $r^2 = 0.773$, $p < 0.001$.

density and property value, all of which both depend upon footfall, and generate footfall. This process of emergence in which spatial configuration plays a key role is called the 'movement economy' (Hillier and Penn, 1992).

The properties of 'natural movement' and its effects in constructing the 'movement economy' are both essentially generic in that they arise directly out of the nature of systems of spaces as these affect patterns of movement. In this sense, architecture displays a number of properties that might be thought of as 'generically functional' (Hillier, 1996). There are a number of other 'generic function' properties in addition to those mentioned already. For example, within buildings we commonly find at least two types of movement: all-to-all movement or the background 'natural movement' of a building must be thought of separately to in-to-out movement – that which starts or finishes at building entrances. This property, which either leads to buildings in which visitors entering or leaving a building are brought together with inhabitants moving around within it if the two systems use the same set of spaces, or in which they are kept apart if they prioritise different spaces, explains the diversity of interfaces constructed between inhabitants and visitors in different buildings. Lastly, buildings vary in terms of the degree to which they are 'intelligible' or maze like. This seems again to be a property of generic function, and can be shown to be a direct consequence of the spatial configuration of the building plan. Intelligibility, and its close counterpart in the degree to which patterns of human use and movement are 'predictable' from the immediately visible properties of the spatial environment, carries strong social implications. If we cannot understand where we are, or cannot predict the likely presence of others, we are deprived of the ability to act intentionally, and effectively have our autonomy removed so far as socio-economic life is concerned.

The ability to act autonomously and with intention is a fundamental requirement for a social agent. Here then is the central thesis of space syntax research: buildings and cities construct, amongst other things, intelligible patterns of space in which the presence of other people and groups is to some degree predictable. It is these patterns of potential presence, and their systematic patterning, albeit as a probabilistic field, that Hanson calls the 'virtual community' (Hanson and Hillier, 1987). It is the virtual community that architects manipulate and form as they design building and urban plans.

It is worth discussing two of the main implications to arise from this thesis and the empirical findings upon which it is based: first, that an important role played by architecture is in the generation and reproduction of social structure; and second, the implications this holds for the relations between architectural practice and research, and the different roles of design practitioner, historian and critical theorist in the way that innovation takes place in this domain. Taken together these findings cast the role of architectural research into a particularly strategic place with respect to social and cultural innovation.

Conclusion

The linking of architectural design directly to the evolution of social structures in the way implied by the research above means that we must take seriously some of the problems faced by social theory. Sociology faces a conundrum: how is it possible for social forms to both evolve rapidly (as no doubt they do) and to show remarkable stability over time frames longer than the lives of the individuals that compose them? The proposition that has emerged from the space syntax research field is that the built environment we construct around ourselves may play an important role both in helping to conserve and reproduce social forms, and in providing the medium for the generation of new social forms. In fact, this axis from the 'conservative' to the 'generative' social or organisational form seems to map onto quite different properties of spatial layout. Generative forms tend to be ringy and shallow, while conservative forms tend towards depth and tree structures. The explanation for this is simple. One of the main effects of spatial structure in the built environment is to define patterns of movement. A side-effect of these is on patterns of encounter and co-presence between different groups of people as they move around an environment. As people within a building move from place to place to fulfil their roles in the organisation, the configuration of the circulation network coupled to the locations of specific functional spaces in that network (the origins and destinations) generates a pattern of movement. In a tree-like structure there is only one sensible route between any two locations, and the deeper that structure on average, the less likely that any two sets of routes will share a common set of spaces. This means that a random meeting between people on different missions (and so those fulfilling different social or organisational roles) is reduced in this kind of structure. Conversely, in ringy and shallow structures there is both a greater choice of routes and greater mixing between those on different routes from place to place. These properties are a natural consequence of the layout of the environment and can go a long way to explain how different organisational cultures might be inscribed in, and partly reproduced by, the spatial design of the built environment.

The idea that one of architecture's functions is to construct and reproduce social structures has deep implications not only for architectural practice, but also for architectural research. Ultimately, if the role of research is to contribute to the body of knowledge, then a distinction must be drawn between two kinds of knowledge: scientific and social. Scientific knowledge is explicit, linguistic or mathematical in its expression. It allows 'what if' questions to be addressed and forms the basis for goal-directed design practice. Social knowledge however is most often tacit and incorporated in the accepted behaviours and norms of a society. It is generated by communities of practice through exploratory design processes. It seems that the field of architectural research contributes to both types of knowledge; the former in building sciences, technological and engineering research; the latter through practice. There are, however, two continual processes through which the boundaries between these types of knowledge are negotiated. First, the critical theorist draws from practice and brings into the realm of explicit discussion the works of the practitioner, while the historian sets these acts into their wider social, political and technological context. Second, there is an underlying move towards explicit and scientific explanation of broadly social and cultural processes. The space syntax research field described earlier gives a clear example of this, however there are many other dimensions in architectural research which follow a similar trend, moving our understanding from tacit to explicit. As each of these fields of knowledge matures it gives rise to new science-based divisions of labour in practice – essentially new branches of engineering – in which explicit knowledge allows 'what if' questions to be answered and enables more goal-directed design processes.

Does this mean that in the long run, the programme of architectural research will reduce everything to explicit scientific understanding and goal-directed design? The argument presented in this chapter suggests not. The strongly emergent nature of this field suggests that although explicit knowledge will grow over time, neither the brief nor the range of design possibilities will ever be fully open to definition at the outset. Certainly, the emergent coupling between these two suggests that innovation must take place continuously and that the situation in the future will never be determined to the point at which purely goal-directed design strategies are feasible, even if these were thought desirable.

Note

1 In a paper for the RIBA Intelligence Unit on the Structure of the Architectural Profession, c., 1970, I have been unable to trace the original.

References

Bustard, W. (1999) Space, evolution, and function in the houses of Chaco Canyon, *Environment and Planning B*, 26(2), 219–240.

Hanson, J. (1999) *Decoding Homes and Houses*, Cambridge University Press, Cambridge.

Hanson, J. and Hillier, B. (1987) The architecture of community: Some new proposals on the social consequences of architectural planning decisions, *Architecture et Comportment/Architecture and Behavior*, 3(3), 251–273.

Hillier, B. (1996) *Space is the Machine*, Cambridge University Press, Cambridge.

Hillier, B. and Hanson, J. (1984) *The Social Logic of Space*, Cambridge University Press, Cambridge.

Hillier, B. and Penn, A. (1992) Dense civilisations: The shape of cities in the 21st century, *Applied Energy*, 43, 41–66.

Hillier, B., Hanson, J. and Graham, H. (1987) Ideas are in things: An application of the space syntax method to discovering house genotypes, *Environment and Planning B: Planning and Design*, 14(4), 363–385.

Hillier, B., Hanson, J., Peponis, J., Hudson, J. and Burdett, R. (1983) Space syntax, *Architects Journal*, 178(48), 67–75.

Hillier, B., Musgrove, J. and O'Sullivan, P. (1972) Knowledge and design, in Mitchell, J.W. (ed.) *Environmental Design Research and Practice*, University of California Press, Los Angeles.

Hillier, B., Penn, A., Hanson, J., Grajewski, T. and Xu, J. (1993) Natural movement: Or, configuration and attraction in urban pedestrian movement, *Environment and Planning B: Planning and Design*, 20(1), 29–66.

Jamous, H. and Pelouille, B. (1970) Professions or self perpetuating system?, in Jackson, J.A. (ed.) *Professions and Professionalisation*, Cambridge University Press, Cambridge.

Chapter Three
Legal research

Paul Chynoweth

Introduction

Legal researchers have always struggled to explain the nature of their activities to colleagues in other disciplines. If Becher's (1981, p. 111) work continues to represent an accurate account of how academic lawyers are viewed by their peers they have much work still to do in this respect. He found that they were regarded as 'not really academic ... arcane, distant and alien: an appendage to the academic world ... vociferous, untrustworthy, immoral, narrow and arrogant'. Their research fared no better, being dismissed as '... unexciting, uncreative, and comprising a series of intellectual puzzles scattered among large areas of description'.

This chapter therefore presents a welcome opportunity to explain the actual nature of legal research (or 'legal scholarship' as it is more usually described) to researchers from the other component disciplines within the built environment. The built environment is usually considered to be an interdisciplinary (or, at the very least, a multidisciplinary) field linking the disciplines of management, economics, law, technology and design (Chynoweth, 2006). The field as a whole can benefit from an improved understanding of each of its component disciplines, and from the greatest possible involvement of each of these in its collective research agendas. The current chapter aims to assist this process in the context of the law discipline. Specifically, it attempts to describe the nature of research within that discipline by reference to the epistemological, methodological and cultural features which distinguish it from other forms of built environment research.

The epistemology of legal scholarship

Legal research styles

There is a dearth of theoretical literature on the nature of legal scholarship and a consequent lack of awareness about what legal scholars actually do. Although there is a tradition of theoretical scholarship (or 'jurisprudence') within the law, this tends to address abstract philosophical questions about the nature of law itself. Many lawyers would recognise Bix's (2003) description of jurisprudence as 'theorists talking past each other' and Murphy and Roberts (1987, p. 682) describe its spectacular lack of contribution to the wider discipline in the following terms:

> legal theory has failed to provide any significant explanation or justification of what academic lawyers do (as is normally demanded of the theoretical component of a discipline) and thus of what academic law is or might be.

Nevertheless, in a very different context, Arthurs (1983, pp. 63–71) proposed a useful taxonomy of legal research styles in his report on legal education and research

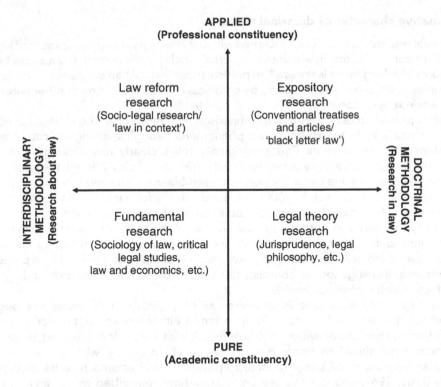

APPLIED
(Professional constituency)

INTERDISCIPLINARY
METHODOLOGY
(Research about law)

DOCTRINAL
METHODOLOGY
(Research in law)

Law reform
research
(Socio-legal research/
'law in context')

Expository
research
(Conventional treatises
and articles/
'black letter law')

Fundamental
research
(Sociology of law, critical
legal studies,
law and economics, etc.)

Legal theory
research
(Jurisprudence, legal
philosophy, etc.)

PURE
(Academic constituency)

Figure 3.1 Legal research styles (Arthurs, 1983).

in Canada. This has informed the analysis in this chapter and is represented as a matrix in Figure 3.1. It will be seen that the vertical axis of the matrix represents the familiar distinction between pure research which is undertaken for a predominantly academic constituency, and applied work which generally serves the professional needs of practitioners and policy makers. However, in the present context, the more interesting distinction is that between doctrinal and interdisciplinary research which is represented by the horizontal axis.

Doctrinal legal research

Doctrinal research (on the right in Figure 3.1) is concerned with the formulation of legal 'doctrines' through the analysis of legal rules. Within the common law jurisdictions legal rules are to be found within statutes and cases (the sources of law) but it is important to appreciate that they cannot, in themselves, provide a complete statement of the law in any given situation. This can only be ascertained by applying the relevant legal rules to the particular facts of the situation under consideration.

As will be discussed below in the section on methodology, deciding on which rules to apply in a particular situation is made easier by the existence of legal doctrines (e.g., the doctrine of consideration within the law of contract). These are systematic formulations of the law in particular contexts. They clarify ambiguities within rules, place them in a logical and coherent structure and describe their relationship to other rules. The methods of doctrinal research are characterised by the study of legal texts and, for this reason, it is often described colloquially as 'black-letter law'.

Normative character of doctrinal research

Doctrinal research is therefore concerned with the discovery and development of legal doctrines for publication in textbooks or journal articles and its research questions take the form of asking 'what is the law?' in particular contexts. At an epistemological level this differs from the questions asked by empirical investigators in most other areas of built environment research.

This is perhaps most obvious in a comparison with research in the natural sciences which typically seeks to explain natural phenomena through studying the causal relationships between variables. Epistemologically, this is clearly very different from the interpretive, qualitative analysis required by doctrinal research. Although the interpretive nature of the process bears a superficial resemblance to the *verstehen* tradition of the social sciences (Schwandt, 2000), there are actually fundamental epistemological differences between doctrinal analysis and all styles of scientific research.

Scientific research, in both the natural and social sciences, relies on the collection of empirical data, either as a basis for its theories, or as a means of testing them. In either case, therefore, the validity of the research findings is determined by a process of empirical investigation. In contrast, the validity of doctrinal research findings is unaffected by the empirical world.

Legal rules are *normative* in character as they dictate how individuals *ought* to behave (Kelsen, 1967). They make no attempt either to explain, predict, or even to understand human behaviour. Their sole function is to prescribe it. In short, doctrinal research is not therefore research *about* law at all. In asking 'what is the law?' it takes an internal, participant-orientated epistemological approach to its object of study (Hart, 1961) and, for this reason, is sometimes described as research *in* law (Arthurs, 1983).

As will be described below, the actual process of analysis by which doctrines are formulated owes more to the subjective, argument-based methodologies of the humanities than to the more detached data-based analysis of the natural and social sciences. The normative character of the law also means that the validity of doctrinal research must inevitably rest upon developing a consensus within the scholastic community, rather than on an appeal to any external reality.

Interdisciplinary research

In practice, even doctrinal analysis usually makes at least some reference to other, external, factors as well as seeking answers that are consistent with the existing body of rules. For example, an uncertain or ambiguous legal ruling can often be more easily interpreted when viewed in its proper historical or social context, or when the interpreter has an adequate understanding of the industry or technology to which it relates. As the researcher begins to take these extraneous matters into account, the enquiry begins to move leftwards along the horizontal axis in Figure 3.1, in the direction of interdisciplinary research.

There comes a point, towards the left-hand side of the matrix, when the epistemological nature of the research changes from that of internal enquiry into the meaning of the law to that of external enquiry into the law as a social entity. This might involve, for example, an evaluation of the effectiveness of a particular piece of legislation in achieving particular social goals or an examination of the extent to which it is being complied with.

In taking an external view of the law, each of these examples could be described as research *about* law rather than research *in* law. As one continues to move leftwards along the axis one encounters a greater willingness to embrace the epistemologies and methodologies of the social sciences.

Pure and applied legal research

Finally, let us return to the distinction between pure and applied legal research represented by the vertical axis in Figure 3.1. Within the context of interdisciplinary legal research (to the left of Figure 3.1) this distinction, in one sense, simply represents that between pure academic knowledge about the operation of the law (at the bottom of the diagram), and knowledge of the same kind which has been produced with a particular purpose in mind (at the top). The purpose of the latter will generally be to facilitate a future change, either in the law itself, or in the manner of its administration. Arthurs (1983) therefore describes this latter category of research as 'law reform research'. The terms 'law in context' and, increasingly, 'socio-legal research' are more often used in the UK. He distinguishes these forms of research from the production of pure, academic knowledge which he refers to as 'fundamental research'.

In fact, there is also a strong correlation between pure, fundamental research and the willingness (indeed, the motivation) of researchers in these areas to question not simply the operation of law, but also its underlying philosophical, moral, economic and political assumptions. Research of this nature takes many forms but would include the Sociology of Law as well as the (left wing) Critical Legal Studies and (right wing) Law and Economics movements.

The applied form of doctrinal research (to the right of Figure 3.1) is concerned with the systematic presentation and explanation of particular legal doctrines and is therefore referred to as the 'expository' tradition in legal research. This form of scholarship has always been the dominant form of academic legal research (Card, 2002) and has an important role to play in the development of legal doctrines through the publication of conventional legal treatises, articles and textbooks.

When doctrinal research is undertaken in its pure form it is variously described as legal theory, jurisprudence, or (occasionally) legal philosophy. The limitations of this form of research in defining the nature of law as an academic discipline have already been noted. Nevertheless, although rarely used as a practical basis for legal analysis, it does provide insights into the nature of the legal methodologies actually employed by lawyers and legal scholars and this will be considered in the next section.

In search of a methodology

Significance of the doctrinal tradition

The dominance of the expository, doctrinal tradition in legal scholarship has already been noted. However, it is important to understand that this is not simply a single, isolated category of scholarship. Some element of doctrinal analysis will be found in all but the most radical forms of legal research.

For example, although law reform research appears as a separate category within Figure 3.1, its practitioners emphasise the importance of traditional legal analysis within their socio-legal work (Cownie, 2004, p. 55). Indeed, even within socio-legal studies, it was once suggested that social scientists should be regarded as 'intellectual sub-contractors' who should be kept 'on tap, not on top' (Campbell and Wiles, 1976). Doctrinal analysis therefore remains the defining characteristics of academic legal research and the account which follows represents an attempt to describe the nature of the methodologies employed within it.

The scale of the task is more daunting than readers may imagine. As already noted, the process of doctrinal analysis is more at home within the humanities than the sciences. Its approach involves the development of scholastic arguments for subsequent criticism and reworking by other scholars, rather than any attempt to deliver results which purport to be definitive and final. Any 'methodologies' in this type of research are therefore employed subconsciously by scholars (and by practising lawyers) who would most usually consider themselves to be involved in an exercise in logic and common sense rather than in the formal application of a methodology as understood by researchers in the scientific disciplines.

Doctrinal research methodology and deductive reasoning

The starting point is to recognise that there is no fundamental distinction between the process of academic doctrinal analysis and the legal analysis undertaken by practising lawyers or judges. As already described, the aim, in each case, is to answer the question 'what is the law?' in a particular situation. In the case of practising lawyers or judges this will be a real and well-defined situation requiring an immediate answer to the question. For the legal scholar, the situation, or more likely the class of situations being considered, will be hypothetical and the purpose is to undertake a more in-depth analysis which is capable of informing the deliberations of practitioners and judges in future cases.

In either case, the initial process of applying a rule of law to a factual situation can be understood as an exercise in deductive logic. Most readers will need no explanation of this form of reasoning which, of course, also forms the basis of the scientific method. However, in a legal context, the familiar syllogism, comprising major premise, minor premise and conclusion, takes the following form:

- Major premise – identifies a general rule of law which requires a specified legal outcome when particular facts are present in a situation.
- Minor premise – describes a particular factual situation.
- Conclusion – states whether the rule in the major premise therefore applies to the facts in the minor premise, and whether the specified legal outcome therefore takes effect.

By way of example, in English law, section 108 of the Housing Grants, Construction and Regeneration Act 1996 contains a general rule of law (the major premise) that a party to a construction contract is entitled to refer a dispute under the contract to adjudication. Therefore, where a particular dispute arises in a particular construction contract between a particular employer and a particular contractor (the minor premise) we can conclude, as a matter of deductive logic, that either party is entitled to refer that dispute to adjudication (conclusion).

Open texture of rules

This, of course, is an idealised account of the process of legal reasoning. If the process were as simple, and as mechanistic as this, society would have no need for lawyers, and still less for legal scholarship. In reality, in almost all cases, the deductive model will fail, without further analysis, to produce a definitive answer to the question of what the law is in a given situation.

Legal rules, of necessity, have to be expressed in general terms and were famously described by Hart (1961) as having an 'open texture', and therefore capable of interpretation in more then one sense. In the context of the above example, there has, for instance, been considerable judicial and academic discussion over the meaning of 'dispute' in relation to construction adjudication. There will, therefore, often be an element of doubt as to whether a rule applies to a particular factual situation and this characteristic will, of course, be manipulated by the opposing parties and their lawyers in an attempt to achieve the outcome that is most favourable to their interests.

Although Hart (1961) concluded that judges exercise discretion in these so-called 'hard cases', their decisions are actually based on recognised patterns of reasoning employed within the legal community which are used to supplement the deductive model described above. Lawyers and legal scholars are therefore often able to predict the outcomes of future cases by employing, however subconsciously, the same patterns of reasoning that will eventually be used by the judiciary.

Role of analogy

The most widely used technique is undoubtedly the process of analogical reasoning. In contrast to deductive reasoning, which entails reasoning from a general rule to a specific case, analogy involves a process of reasoning from one specific case to another specific case. In those many situations where it is unclear whether a particular factual situation falls within the ambit of a rule, it can often be helpful to examine apparently similar cases which have previously come before the courts. If, upon examination, the facts of these cases are found to be sufficiently similar to the facts of the subject case then it can be concluded that the facts of the subject case should be treated by the courts in the same way. Most readers will be familiar with this process in the context of the operation of the common law doctrine of precedent.

The decision as to whether a case is sufficiently similar to another is ultimately a subjective one as no two cases are ever completely identical. Judges therefore have considerable scope to distinguish the facts of a subject case from those in an established precedent if they choose not to follow it. Nevertheless, this scope is not unlimited and Bell (1986, p. 48) has highlighted how judicial decision making in these circumstances is constrained by social conventions within the legal community which he describes as the 'rules of legal discourse'. He describes how these 'provide a framework lying outside the power of the reasoner within which he has to operate if his arguments are to count as legal justifications'. Judges are subject to these rules but so, of course, are lawyers and legal scholars who all participate in the same legal discourse, and who all desire their arguments to be taken seriously.

Induction and legal formalism

A third technique involves the use of inductive reasoning which can be described as the reasoning from specific cases to a general rule. This can be of particular assistance when a particular factual situation does not appear to be addressed directly by a legal rule at all and it therefore becomes necessary to 'fill the gap' in the law. As with inductive reasoning in the sciences a general proposition can sometimes be derived from a number of specific instances.

In the case of legal reasoning this involves the recognition of a new general rule which emerges from a number of earlier authorities which are then regarded simply as particular instances of the new rule. *Donoghue v Stevenson* [1932] AC 562 is the best-known example of this technique. Particular instances of negligence had been recognised by the courts for years before the famous snail in the ginger beer case came before the courts. However, it was not until Lord Atkin proposed his now well-known neighbour principle in this case that the tort of negligence was recognised as a more general rule, capable of being applied to novel fact situations which were not already described in the individual authorities then available. Once again, the capacity for developing new rules in this way will be regulated and limited by the recognised rules of legal discourse described above.

A variety of other techniques is available which, like those already described, also allow the available body of legal rules to be marshalled into coherent patterns (or 'doctrines') and applied to new factual situations in an apparently logical and consistent manner. Indeed most legal discourse revolves around the verbal manipulation of the available sources of law, in the belief that the answer to most legal problems can be found in the underlying logic and structure of the rules if only this can be discovered (Smith, 2004). This approach is usually described as legal formalism (Vandevelde, 1996) and, despite numerous academic criticisms of its assumptions (e.g., Fitzpatrick and Hunt, 1987), continues to represent the dominant paradigm within legal practice and within legal scholarship, at least in terms of external appearances.

Indeterminacy and policy judgements

Nevertheless, there is now a widespread recognition that, in some cases, the law cannot be determined with certainty from an analysis of the rules alone. Although judges will justify their decisions by reference to the existing rules (MacCormick, 1994) there is a growing realisation that the rules (in the so-called 'hard cases') can sometimes be used to justify a number of possible, and opposing, legal outcomes. This is, once again, a function of the open texture of legal rules and, where this occurs, the law is said to be indeterminate (Kress, 1989).

If the law is indeterminate, and some cases are decided according to a value judgment made by the judge on the day, there are of course implications for democracy, and for the rule of law. This has unsurprisingly generated criticisms of the political role of the judiciary (e.g., Griffith, 1997), which remains beyond the scope of this chapter. However, the judges' political role is usually described more charitably in terms of making decisions according to 'policy considerations' and this is now widely accepted as a legitimate part of the judicial function.

The challenge for the legal scholar (or practising lawyer) trying to predict the likely outcome of future cases is to understand the nature of the policy considerations that are likely to influence the judiciary. Dworkin's (1977, 1986) influential writings provide a wealth of guidance in this respect and remind us that policy decisions are far from the arbitrary and unpredictable exercise of judicial power that some would suggest. Rather, he argues that legal systems consist of underlying principles, as well as rules, and that judges are bound to follow these when deciding the outcomes of hard cases. As with Bell's (1986) rules of legal discourse described above, these can be seen to provide a constraint on judicial action, and at least some assistance in attempting to anticipate the likely outcome of cases. Bell's (1983) empirical work on policy matters also identifies the particular forms of policy argument used by the courts and this can also assist the scholar in trying to anticipate judicial decision making in this context.

Summary

In summary, therefore, it is probably incorrect to describe the process of legal analysis as being dictated by a 'methodology', at least in the sense in which that term is used in the sciences. The process involves an exercise in reasoning and a variety of techniques are used, often at a subconscious level, with the aim of constructing an argument which is convincing according to accepted, and instinctive, conventions of discourse within the discipline.

Although the discourse is apparently conducted according to formalistic conventions it is also influenced by shared value (or policy) judgments which often remain unspoken. The 'methods' employed in legal scholarship are therefore neither consciously learned, nor consciously employed as is the case with scientific methods. The skills and conventions of legal analysis are instead learned at an instinctive level through exposure to the process, and they are then employed on the same basis in the development of legal argument. In much the same way that the use of an explicit methodology confers legitimacy in scientific research, credibility within legal scholarship is therefore dependent on the researcher's work demonstrating an understanding and adherence to the accepted conventions and norms of its discourse.

The cultural dimension

Disciplinary spectrum

This lack of a formal research methodology, and the reliance on analysis and the development of argument within a prevailing academic discourse, is of course a particular feature of the arts and humanities family of disciplines to which law belongs. This places law at the 'soft' end of the familiar disciplinary spectrum. Using the well-known Biglan (1973) disciplinary model (illustrated in Figure 3.2) it can be seen that (in common with design), law differs from the dominant built environment research specialisms in this respect. Unlike law and design, the disciplines of technology, economics and management all belong either to the natural, or to the social sciences.

The science/arts & humanities distinction reflects genuine epistemological and methodological differences between the families of disciplines about the nature of

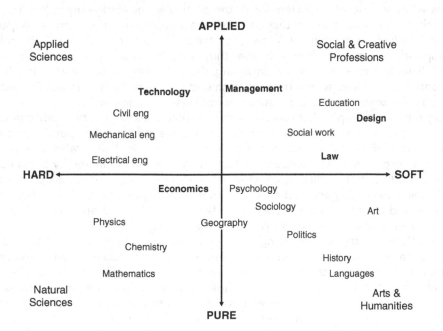

Figure 3.2 Disciplinary model (Biglan, 1973).

knowledge, and about the manner of its production. Becher (1987) has described knowledge production in the sciences in terms of the cumulative and piecemeal accumulation of individual segments of knowledge which, over time, contribute to a comprehensive explanation of particular phenomena. He contrasts this with humanities disciplines like law. These, he describes, as being concerned with the organic development of knowledge through an ongoing process of reiterative enquiry. They address multifaceted, rather than discrete, problems and attempt, not to explain the individual components of phenomena, but to develop a holistic understanding of their overall complexity.

The dominance of the scientific disciplines within the built environment inevitably influences prevailing views about knowledge and knowledge production within the field. Indeed, the language of built environment research is often dominated by the rhetoric of the social sciences in particular. This is characterised by a concern with the traditional social science methodologies (see, e.g., Fellows and Liu, 2003) and with an emphasis on empirical investigations rather than the development of theoretical perspectives (Betts and Lansley, 1993; Brandon, 2002).

Cultural challenges

The epistemological and methodological differences between legal scholarship and most other built environment research styles also generate cultural differences between the two. These produce expectations regarding the external appearance of academic research within the field which legal scholars often struggle to satisfy. These may relate to expectations about the form and appearance of research outputs, about the process which is undertaken in generating the research, and about the more general behavioural characteristics of researchers within the field.

In their seminal work, *Academic Tribes and Territories*, Becher and Trowler (2001) have demonstrated how individual academic communities (tribes) develop cultural norms which are closely associated with the particular knowledge areas (territories) which they inhabit. In particular they demonstrate a close correlation between Biglan's (1973) hard/soft continuum of knowledge types (illustrated in Figure 3.2) and a corresponding continuum between what they describe as urban and rural research styles. Scientific research culture (including the prevailing culture within the built environment) conforms to an urban research pattern whilst the humanities (including law) typically exhibit the characteristics of a rural research community.

They find, for example, that urban research communities like the built environment focus on narrower and more short-term research topics, are more competitive and are more influenced by the availability of external funding than their rural counterparts. They also describe a greater tendency for urban areas to be dominated by charismatic research leaders (the so-called 'research stars') than rural areas. Urban research is faster moving and more gregarious than that within rural environments and is therefore characterised by more networks, a higher level of conference attendance and an increased incidence of team working than in rural settings.

The different patterns of working are also reflected in publication patterns and styles. Urban communities produce large numbers of short articles, often by multiple authors, whilst the outputs from rural communities like law are likely to be substantial, but less frequent, and authored by a single researcher. The gregarious teams of researchers in frenetic urban environments can therefore easily overlook their more solitary, and less visible, counterparts in rural fields. The danger for those operating in the rural subjects like law is that their lack of visibility can be mistaken for lack of activity.

More fundamentally, as illustrated by the quotations at the start of this chapter, cultural differences can sometimes obscure the academic merits of doctrinal work from those belonging to different disciplinary traditions. As a consequence, legal scholars' experiences of peer review within the built environment have not always been happy one's. Their work can all too easily be dismissed as lacking a methodology, as being based only on opinion, or even as being 'not research' by peers operating within a scientific, rather than a humanities paradigm.

Conclusion

The chapter has shown that the normative process of doctrinal analysis is the defining characteristic of most legal scholarship. It has demonstrated how this places it within the humanities' tradition with corresponding methodologies and cultural norms. As the built environment research community operates overwhelmingly within a scientific paradigm it embraces different methodologies and cultural norms from those traditionally associated with legal scholarship with consequent difficulties for communication.

In common with other humanities' disciplines, most legal scholarship is not concerned with empirical investigation, but with the analysis and manipulation of theoretical concepts. The methodologies employed therefore differ from those of the sciences and are probably more accurately categorised, in social science terms, as techniques of qualitative analysis. As has been seen, deductive and inductive logic, the use of analogical reasoning and policy analysis all feature strongly within this process.

Crucially however, as the process is one of analysis rather than data collection, no purpose would be served by including a methodology section within a doctrinal research publication and one is never likely to find one. This is perhaps the most striking difference between the appearance of research outputs in the two traditions, and the one which has historically caused most difficulty for legal scholars when subject to peer review by other built environment researchers.

This chapter began by highlighting the failure of the legal research community to adequately explain itself to its peers in other disciplines and, in this sense, it can hardly complain if those peers then judge it by standards other than its own. Communication between disciplines is one of the great challenges to achieving genuine interdisciplinary rigour and that challenge is never greater then when trying to bridge the gulf between the humanities and the sciences.

Nevertheless, it is surely incumbent on all of us within the built environment research community to do precisely that. This involves developing at least an awareness of practices within the field's various disciplines. But it also involves a willingness to reflect upon our own previously unquestioned assumptions about the practices in our own discipline, and to articulate these for the benefit of others within the field. It is hoped that the above account might make some contribution to this process by increasing understanding (perhaps amongst legal scholars as well as others) about the nature of legal research, and about how it differs from other research within the built environment.

References

Arthurs, H.W. (1983) *Law and Learning: Report to the Social Sciences and Humanities Research Council of Canada by the Consultative Group on Research and Education in Law*, Information Division, Social Sciences and Humanities Research Council of Canada, Ottawa.

Becher, T. (1981) Towards a definition of disciplinary cultures, *Studies in Higher Education*, 6, 109–122.

Becher, T. (1987) The disciplinary shaping of the profession, in Clark, B.R. (ed.) *The Academic Profession*, University of California Press, Berkeley, California.

Becher, T. and Trowler, P.R. (2001) *Academic Tribes and Territories*, SRHE and Open University Press, Buckingham.

Bell, J. (1983) *Policy Arguments in Judicial Decisions*, Clarendon Press, Oxford.

Bell, J. (1986) The acceptability of legal arguments, in MacCormick, N. and Birks, P. (eds) *The Legal Mind: Essays for Tony Honoré*, Clarendon Press, Oxford.

Betts, M. and Lansley, P. (1993) Construction management and economics: A review of the first ten years, *Construction Management and Economics*, 11(4), 221–245.

Biglan, A. (1973) The characteristics of subject matters in different academic areas, *Journal of Applied Psychology*, 57(3), 195–203.

Bix, B. (2003) *Jurisprudence: Theory and Context* (3rd Edition), Sweet & Maxwell, London.

Brandon, P. (2002) *Overview Report on 2001 Research Assessment Exercise, Panel No. 33: Built Environment*. Higher Education & Research Opportunities in the United Kingdom, available at http://admin.hero.ac.uk/rae/overview/docs/UoA33.doc

Campbell, C.M. and Wiles, P. (1976) The study of law in society in Britain, *Law and Society Review*, 547–578.

Card, R. (2002) The legal scholar, *The Reporter: Newsletter of the Society of Legal Scholars*, 25, 5–12.

Chynoweth, P. (2006) The built environment interdiscipline: A theoretical model for decision makers in research and teaching, *Proceedings of the International Conference on Building Education and Research* (CIB W89 BEAR 2006), 10–13 April 2006, Hong Kong Polytechnic University, Hong Kong, People's Republic of China.

Cownie, F. (2004) *Legal Academics: Culture and Identities*, Hart Publishing, Oxford and Portland, Oregon.

Dworkin, R. (1977) *Taking Rights Seriously*, Duckworth, London.

Dworkin, R. (1986) *Law's Empire*, Harvard University Press, Cambridge, Massachusetts.

Fellows, R. and Liu, A. (2003) *Research Methods for Construction* (2nd Edition), Blackwell Publishing, Oxford.

Fitzpatrick, P. and Hunt, A. (eds) (1987) *Critical Legal Studies*, Basil Blackwell, Oxford.

Griffith, J.A.G. (1997) *The Politics of the Judiciary* (5th Edition), Fontana Press, London.

Hart, H.L.A. (1961) *The Concept of Law*, Clarendon Press, Oxford.

Kelsen, H. (1967) in Knight, M. (trans) *The Pure Theory of Law*, University of California Press, Berkeley, California.

Kress, K. (1989) Legal indeterminacy, *California Law Review*, 77(235), 283–337.

MacCormick, N. (1994) *Legal Reasoning and Legal Theory*, Clarendon Press, Oxford.

Murphy, W.T. and Roberts, S. (1987) Introduction (to the Special Issue on Legal Scholarship), *Modern Law Review*, 50(6), 677–687.

Schwandt, T.A. (2000) Three epistemological stances for qualitative inquiry: Interpretivism, hermeneutics and social constructionism, in Denzin, N.K. and Lincoln, Y.S. (eds) *Handbook of Qualitative Research* (2nd Edition), Sage Publications, Thousand Oaks, London and New Delhi.

Smith, S.D. (2004) *Law's Quandary*, Harvard University Press, Cambridge, Massachusetts and London.

Vandevelde, K.J. (1996) *Thinking Like a Lawyer: An Introduction to Legal Reasoning*, Westview Press, Boulder, Colorado.

Chapter Four

Feminist research

Pat Morton and Sara Wilkinson

Introduction

This chapter explains feminist research paradigms and demonstrates their relevance and application in built environment disciplines. The chapter provides readers with an understanding of the term 'feminist research' and the techniques associated with this approach. After an initial explanation, we demonstrate how feminist research fits with broader social theories and briefly how it developed. The key concepts and arguments around feminist epistemologies are then identified. We then explain techniques used by researchers adopting feminist approaches such as participatory research and various qualitative methods. The remainder of the chapter provides examples of how researchers can use this approach, using examples within the built environment. There is an extensive amount of work in this field and this chapter is limited to an overview of key aspects and, therefore, readers are provided with a list of further useful readings which will extend and deepen their knowledge.

What is feminist research?

What makes research feminist? One answer is that it is research done by, for, and about women (Burns and Walker, 2005, p. 66). Although no sole definition of feminist research is in existence, and in some minds a universal definition is not desirable, many feminist researchers recognise fundamental characteristics. These characteristics differentiate feminist research from traditional social science research; that it is research that focuses on gender or that studies women. Feminist research is more than method because it raises questions about ontology (one's view of the world and how this shapes what can be known about the world and what it means to be human) and epistemology (what counts as knowledge and ways of knowing) (Code, 2000). For Reinharz (1992, pp. 243–244) feminist research is a 'perspective' and not a method. It is the types of questions, methodologies, knowledge and purpose brought to the research process that makes feminist research distinctive and unique.

Historically, feminist research was informed by women's toils against oppression and adherence to feminist values and beliefs. However, a fundamental tenet is that feminism is a:

> belief that women and men are inherently of equal worth. Because most societies privilege men as a group, social movements are necessary to achieve equality between women and men. (Freedman, 2002)

There are three key features of feminist research. Firstly, the research is characterised by objectives to build new knowledge and to achieve social change. Secondly, feminist research is based on the values and beliefs of feminism and includes feminism within the process, for example, to centre on the meanings women give to their world, and at the same time accept that research is often undertaken in patriarchal organisations or environments. With this approach, feminist values inform the research, starting from the selection of the research issue to data presentation. Adopting a feminist paradigm invokes a structure which leads the researcher's decision making. A third feature is the diversity of feminist research which is both interdisciplinary and trans-disciplinary. Though feminist research does adopt various methodologies it is redefined continuously by the concerns of women coming from very different viewpoints or perspectives. Consequently, it requires that issues such as the empower-ment of women, including those traditionally excluded, are attended to as well as issues of diversity, racism and democratic decision making.

Social change during the 1980s and 1990s has increased the amount of legislation prohibiting discrimination on the basis of gender and women now play an equivalent part in working life. However, women are still experiencing barriers to progression in the workplace, inequality of opportunity and discrimination. Furthermore, the expansion of tertiary education to the masses since the 1990s has led higher numbers of women to participate in higher education with some increase in enrolment in non-traditional courses such as those covered by the built environment. Consequently, awareness has been raised amongst academics (Turrell et al., 2002; Turrell and Wilkinson, 2005) about the experiences women are having both in education, academia and the workplace, which has become the focus of research. There is a growing body of research into women's experiences in built environment disciplines (Greed, 1991, 1999; Turrell and Wilkinson, 2005; Lingard and Sublet, 2002), which adopt feminist paradigms and can be understood more fully with an understanding of the paradigm. A third area of expanded resources is the European Union and UK Government-funded recognition of employment sectors that have failed to make progress in women's equality and participation to the detriment of the economy. The UK Resource Centre for Women in Science, Engineering and Technology (UKRC) is an example (www.ukrc4setwomen.org). The UKRC was set up in 2004 by the Department of Trade and Industry to provide a source of information and advice to all those who aim to improve the recruitment, retention and progression of women in science, engineering and technology (SET), including the built environment. These schemes are predicated to work with employers, professional institutions, policy makers and other stakeholders as well as women to bring about change. Feminist research informs strategic developments and appropriate support mechanisms and, of course, with their goal to change social conventions as a key characteristic, they conform to the paradigm.

It should also be noted that feminist research methods have characteristics that can be imported into other research paradigms, such as the focus on deep, rich qualitative data. For example, standpoint theory can be applied in research on organizational culture and issues relating to ethnicity and race or on environmental issues – all of which feature highly in current built environment research concerns. In summary, there is much to be learned from an appreciation and understanding of feminist research.

Locating feminism in the social sciences

Generally, methodology is the study of methods and practices employed in research which involve the gathering of evidence in the process of knowledge and theory formation. The so-called traditional methodology is based on liberal-positivist epistemology using empiricism, objectivity and rationalism as fundamental principles. Feminist research methodology critiques the theoretical principles of the traditional approach from numerous perspectives. Epistemology, or the theory of knowledge, is the branch of Western philosophy that studies the nature and scope of knowledge and belief. Feminist epistemology asks 'whose knowledge are we talking about?' and feminist methodology asks 'how should we go about producing knowledge?' (Code, 2000).

Traditional approaches have adopted androcentric assumptions in the design and application of research projects, and according to feminist researchers, relied on the over-generalisation of research based on male-only samples. To clarify, androcentric or androcentrism is a world view that is male centred. Given that approximately half the population is female this has strengthened the call for feminist approaches. However, feminist research is, relatively speaking, a recent addition to the lexicon of research, it is also very dynamic with various philosophical and ethical debates under continuous review (Denzin and Lincoln, 2000).

To comprehend feminist critiques a review of positivist research methodology is needed (Code, 2000). The foundations and requirements for the traditional research process are based on the assumption of the existence of an objective reality that can be logically and rationally discovered through observation. Here the process is led by the construction of hypotheses and the operation of concepts which are tested against the evidence collected. Researchers must maintain objectivity to determine that their research is not biased, and therefore undermined, by personal values. Consequently there are many guidelines in the process of theory development, research design, data collection and analysis to ensure research is not compromised by subjectivity. Positivism makes some assumptions such as knowledge exists outside the lived experience of the objects of study and that truth is revealed through objective rational review of evidence. The objects of study play a limited role in the research process, being unable to achieve objectivity. A final characteristic of the approach is that traditional research often desires quantification of data; the most popular method is the survey. Data are aggregated and summarised and the use of statistical techniques is employed to reveal causal relationships across variables. However, some feminist researchers do not accept this model and question, amongst other things, whether an objective truth can be rationally determined through empirical observation (Stanley and Wise, 1993).

Although it is acknowledged that feminists may not concur how to define feminist research, there is some agreement about the epistemological grounding of the research process. In 1986, Cook and Fonow identified five epistemological principles in feminist methodology, which can be seen as enduring themes, and are as follows:

(1) Women and gender are the focal point of analysis.
(2) The rejection of subject and object.
(3) The importance of consciousness raising.
(4) A concern with ethics.
(5) An intention to empower, alter power relations and inequality for women.

Feminist research varies from traditional or 'scientific' research, primarily because it seeks to extract the power imbalance between researcher and subject and consequently it is politically driven as it aims to alter social inequality. Feminist research thus commences with the standpoints and experiences of women, concepts which are developed later in this chapter. A wide range of methods, both qualitative and quantitative, are available to feminist researchers, however, not surprisingly, there is considerable debate about the appropriateness of these methods (Millen, 1997; Reinharz, 1992). One argument is that rather than concentrating on which research method is best, it is more appropriate to let the research setting and aim lead the selection of tools and techniques (Greaves *et al.*, 1995, p. 334; Reinharz, 1992). As a result, like traditional research paradigms, there is no one method or strategy for feminist research (Bell *et al.*, 1993).

Locating the feminist researcher

As stated previously, there is a debate about whether feminist research is a methodology. Perspective is a cornerstone of feminist research and epistemology conceives of 'knowers' as situated in particular relations to what is known and to other knowers. What is known, and the way that it is known, reflects the perspective of the knower. In other words, consider how people can comprehend the same object in different ways that reflect where they stand in relation to it. One perspective is physical embodiment. People experience the world by using their bodies, which have diverse constitutions and are differently positioned in space and time (Bordo, 1987; Young, 1990). Then there is *first-person* and *third-person* knowledge. People have first-person experience of their own physical and mental circumstances, providing direct knowledge about what it is like to live through these states. Third parties may know these circumstances by interpretation, imagination or through written records. For example, it is one thing to know what gender harassment is, and how to identify it in a case described in third-person terms. It is quite another thing to realise '*I* am being harassed on the basis of gender'.

Then, of course, people represent objects in relation to their emotions, attitudes and interests (see the work of Harding (1986, 1991, 1993, 1998), Gilligan (1993), Diamond (1991), Jaggar (1989) and Keller (1986)). For example, a computer hacker sees a password as a frustrating obstruction, whereas the computer owner perceives the password as security and protection. Another aspect is people's personal knowledge of others. Everyone has different knowledge based on their relationships to others. Sometimes this is tacitly known, like the knowledge it takes to get a joke; it is an interpretive skill. Therefore, as people behave differently towards others, and others interpret their behaviour differently, depending on their personal relationships, what others know of them depends on these relationships. Therefore, within feminist research it can be posited that a female researcher who has experienced the built environment is better placed to undertake research into women's experiences in the built environment.

Some theorists think that men and women have different cognitive styles (Belenky *et al.*, 1997; Gilligan, 1993) and that cognitive styles are gender symbolised (Rooney, 1991). For example, deduction and quantitative cognitive styles are perceived as 'masculine', intuition and qualitative styles are 'feminine'. Consequently this raises questions: does the quest for masculine prestige through masculine methods distort practices of knowledge acquisition (Addelson, 1983)? Are some kinds of research unfairly ignored because of their association with 'feminine' cognitive styles (Keller, 1986)? And are feminine cognitive styles producing knowledge that is inaccessible by 'masculine' means (Duran, 1991)? Clearly, there are enormous ramifications for feminist research in the built environment.

People have different styles of investigation and representation which is drawn, for instance, from background beliefs and worldviews (see MacKinnon, 1999; Harding, 1986; Hubbard, 1990). The researcher's personal history and inside knowledge as a result of life experience, beliefs and background can influence the view of their own research. Leland (in Reinharz, 1992) carried out a study on gay men, drawing from her knowledge as a mother of two sons, one gay and one straight, and was able to draw on inside experience and knowledge to enrich the data. It is these kinds of perspective or situatedness which affect knowledge; see also Kalbfleisch (1995), Addelson (1983) and Nelson (1990).

Feminist research takes into account how the social location of the knower affects what and how she knows. Social location consists of social identities (gender, race, status, ethnicity, caste, kinship, sexual orientation, etc.), social roles and relationships (occupation, political party membership, etc.). Women occupy different social roles that accord them varying powers, duties, goals and interests; furthermore, women have different norms prescribing different virtues, habits, emotions, and skills that are considered appropriate to these roles. As such, women arrive at different subjective identities which can show in various ways. For example, an individual may perceive and accept some attributed identities, affirming the roles connected with them or they may see the social identities as unfair and seek change. The most obvious is gender, and sociologists and feminists distinguish between sex and gender. Sex is the biological difference between male and female whereas gender is what society makes of sexual differences: the roles, norms, and meanings they assign and therefore gender has many facets (Haslanger, 2000). There are gender roles, traits, performance, virtues, norms, behaviours, identity and symbolism (see Butler, 1990).

Ethical considerations

As with traditional research, feminist research is concerned about privacy, consent, confidentiality, deceit, and avoiding harm to those involved in research (Cook and Fonow, 1986; Denzin and Lincoln, 2000). The work of many feminist researchers has led the development of these issues as the research often involves risk groups such as women's health and homelessness. Over time and as with social science research methods in general, ethics has become more complicated and differentiated. Recent issues sit uneasily with older ethical concerns such as informed consent, for example. Traditionally, it was assumed that consent once given did not fade or change over time; now this is contested (Denzin and Lincoln, 2000). Another example is the requirement that the researcher carries out the research in an open and honest manner around data collection, analysis and publication. These issues are now closely linked for feminist researchers with considerations of how and where knowledge is created.

When it comes to participatory research, the key ethical issues are those of voice and account, especially where the research involves women doing research on themselves and on topics of concern to them and their lives. Participatory research confronts researchers with issues such as women's knowledge, representations of women, modes of data collection, analysis, interpretation, writing up, and the relationships between and among collaborating parties. Kleiber and Light's work in 1978 shows their transition from traditional field workers to co-researchers when they studied women's health collectives in Canada. With participatory research there is also the ethical question of ownership of the data and the issue of power remains. Furthermore, with participatory research the issue is raised with interpretation as both researcher and participant interpret data. Previously, researchers interpreted the data they collected without the participants' contributions and this can create concerns where the parties do not concur with the interpretation of data. Belenky et al. (1997) carried out a landmark piece of feminist research on women's higher education experience in 1986, which illustrates many aspects of involvement and subjectivity. The four research-ers explain how they tried to ignore their own theories brought to the project when the women's views diverged and 'we forced ourselves to believe the women' (p. xiii) in order to hand the power back to those being researched.

Standpoint epistemologies

What is standpoint epistemology and why is it relevant to feminist research? Standpoint theories characterise the world from a specific socially situated perspective that can lay a claim to epistemic privilege or authority (Denzin and Lincoln, 2000). It grew from a set of theoretical positions from numerous disciplines but takes the perspective of the marginalised woman's experiences. Leading researchers who developed this area are Harding (1986) and Hartsock (1983).

For standpoint theory to be complete it has to specify the following:

(1)   The *social location* of the privileged perspective.
(2)   The *aspect* of the social location that creates superior knowledge: for example, social role or subjective identity.
(3)   The *scope* of its privilege and what questions or subject matters it alleges a privilege over.
(4)   The *ground* of its privilege: what it is about that aspect that justifies a claim to privilege.
(5)   The *type* of pre-eminence it claims, for example, greater accuracy, or greater ability to show basic truths.
(6)   The *other perspectives* relative to which it claims epistemic superiority.
(7)   Modes of access to that perspective: is occupying the social location required or sufficient for getting access to the perspective?

Many claims to epistemic privilege on behalf of specific perspectives with respect to certain questions are commonplace and uncontroversial. For example, IT technicians are generally in a better position than computer users to know what is wrong with their computers. Experience in performing the social role of the IT technician grounds the IT technician's epistemic privilege, which lays a claim to superior reliability than the opinions of computer users. It is noted, however, that the term standpoint includes a range of theories which are not identical (Denzin and Lincoln, 2000), and that standpoint theories can be controversial when they claim epistemic privilege over socially and politically contested topics on behalf of the perspectives of systematically disadvantaged social groups, relative to the perspectives of the groups that dominate them. Hence, research on gender and the built environment could be considered controversial. The extent of the claimed privilege includes the character, causes and consequences of the social inequalities that define the groups in question. This type of standpoint theory claims three types of epistemic privilege over the standpoint of dominant groups. First, it claims to provide deep over surface knowledge of society: the standpoint of the disadvantaged reveals the fundamental regularities that drive the phenomena in question, whereas the standpoint of the privileged portrays only surface regularities. Second, it claims to offer better-quality knowledge of the modality of surface regularities, and thus greater knowledge of human potentialities. Where the standpoint of the privileged is inclined to represent existing social inequalities as natural and necessary, the standpoint of the disadvantaged correctly represents them as socially contingent, and shows how they could be overcome. Third, it claims to offer a portrayal of the social world in relation to universal human interests. In comparison, the standpoint of the privileged portrays social phenomena relative only to the interests of the privileged, but ideologically misrepresents these interests as corresponding with universal human interests.

Feminist standpoint theory is based on Marxist perceptions of the role of the worker and posits that women, as an oppressed group, can frame their experiences of oppression and see the oppressors and the world more clearly. In reply to the assertion that female experience is an invalid basis for knowledge, feminist standpoint research declares it is a more valid basis for knowledge as it gives access to a broader comprehension of truth via the insight into the oppressor. Some observers have suggested a successor science to existing paradigms, which privileges purported feminine qualities of holistic, integrated, connected knowledge contrasted to the analytical, masculine form of knowledge. Thereby suggesting that there is a feminine notion of knowledge, which is intuitive, emotional, engaged and caring, which has been barred from the advance of ideas about knowledge because women's experiences are not included in that process.

Standpoint theory drawn from ideology that influences ideas and beliefs can also be applied in other areas such as anti-racism, anti-globalisation and environmentalism. The standpoints of the researchers should be clearly stated. The problems can come when a standpoint is not pointed out, but the readers of the research perceive the data as being influenced by a standpoint, that is bias. Hammersley (1995), whose general position is against such theoretical approaches, explores a case where a white middle class teacher doing research found no racism in a city centre school. The findings were criticised widely both on method and substantively. The cultural standpoint of the researcher was seen as a hindrance in getting to the truth. Having considered epistemological issues, the following sections outline various relevant approaches to undertaking feminist research.

Participatory action research

Through the 1990s, participatory action research has evolved and become a methodology for intervention, development and change within the communities and groups such as women in the built environment. The approach is accepted as research which involves all relevant stakeholders in actively examining together a current situation (regarded as problematic) to improve and change it; as such it has much in common with feminist research paradigms. All relevant contexts such as historical, economic, cultural and political are included in a critical review to gain understanding of the context of the problem. It is iterative, described by Morton (2006) as a spiral, whereby participants, with the aid of the researcher, frame and identify the problem(s) and then posit changes to improve the situation, which, in turn, are further reflected on and refined as necessary. Action is taken and participants reflect on the new situation and so it continues until change is achieved. This method is active co-research driven by a democratic approach; placing the participant and researcher in very different positions than traditional methods. The different positions the researcher and participant find themselves in relate to power and ownership of data noted previously.

Clearly, such an approach involves a long-term commitment by all parties to make a successful change. Another issue is that it is difficult to predict outcomes and the challenge, for example of keeping participants involved over a long period, is considerable. Researchers have to be creative, innovative and able to problem solve with this technique. A criticism of the technique by traditionalists is that it is political; participation is empowerment and empowerment is politics and that the goal is to change the position of a 'weaker' group enabled by a 'stronger' person or group (Chambers, 1993). Social change is fundamental to feminist research and an inherent characteristic.

Oral histories and diaries and women's voices

Oral history is a technique used on topical, biographical or autobiographical research projects. It enables the researcher to collect stream of consciousness information related to the research topic and the research participants' feelings about them in an uninhibited setting. A similar method is the use of diaries (Reinharz, 1992; Morton, 2006), which allows participants to record their experiences openly and easily without concerns about grammar or spelling issues. Gilligan (1993), in the second edition of her significant sociological and psychological study, explains (p. xii) the process she went through in listening to women and hearing a different way of speaking from men. Her questions, she said, were about women's perceptions of reality and truth, which get lost when they are reinterpreted in mainstream theories and needed to be received in their own right. Belenky et al. (1997) illustrate how women's voices being heard was a key strand in their methodology as they tried to ensure they were completely open to the women's words describing how they experienced higher education. Greed (1991) talked to women surveyors about their experiences. She reports that she only had to ask them why they went into surveying and 'they were under starter's orders and off, with no stopping them' (Greed, 1991, p. 15). Morton's (2006) work on the experiences of female students on built environment courses in the UK showed how reflection can be used to gather rich, deep data, recording experience and perceptions. Students are asked to reflect openly on a number of issues raised by the researcher in a written form. The reflections were collected by the researcher, interpreted and then discussed further with participants to explore meaning and identify significant aspects that impact on higher education experience.

Can anyone be a feminist researcher?

Feminist research then, is a clear illustration of researchers starting from the viewpoint of their own culture, experience and tradition and carrying out research that really matters to them. This is in contrast with mainstream research where it is stated that personal experience may contaminate the objectivity of the project (Hammersley, 1995). There is a risk here, however, in becoming a convert to the cause and producing naïve research that is clearly biased. The researcher has to maintain a critical approach to the data and demonstrate this throughout the research and within his or her writing up. Stating the position of the researcher as a preface or a postscript, outlining their relationship to the subject matter is an accepted way of declaring interest and approach.

> As feminists we ask the question from a different set of assumptions. We see engineering as a socially constructed profession which is masculine, but we question whether it is inevitably or beneficially so. (Carter and Kirkup, 1990, p. 1)

Can the researcher be a man? The feminist community have been divided over whether men can adopt the role of a feminist (Reinharz, 1992) but there are a few men who profess to label themselves as feminist. The arguments against men say they can never have a women's experience and therefore are unable to understand women. Those who support male researchers say that men do not have to experience being female; they can still contribute to the knowledge area. Because they understand and care about the women's experiences and inequality, they are a valuable element in work to achieve change and equality.

It is likely that the attributes of a researcher adopting a feminist approach will:

- Care about the topic of their research.
- Draw from personal knowledge and experience.
- Want to transform the male tradition or male perspective.
- Be interested in women's views.

And their research will:

- Explore an issue or a topic that relates to the position of the minority group, that is women.
- Take a male-centred topic that needs to be rethought in terms of women's experiences.
- Make visible an aspect of experience that has been unseen.

Feminist research is often driven by its subject matter rather than by its methods (Reinharz, 1992, p. 213), and it is a perspective held rather than a defined method. The construction industry, and the built environment professions within it, are an enduring example of a significantly segregated area of study and work where women are generally a minority and/or where the culture is perceived as masculine. Thus, there are a wide range of topic areas that will benefit from a feminist perspective and the following projects illustrate how researchers in the built environment have been inspired by feminist perspectives.

Greed's (1991) rich study of surveying from a broadly feminist perspective is a landmark within built environment research. Her study was revolutionary within built environment research and broke away from mainstream objectivity by clearly stating a position of wanting change and inserting her own perspective transparently within the data. Her research explored the position of women in surveying, and the likely implications for women and the built environment which they inhabit. She explained her initial approach to the study as listening to both men and women within the profession to find out what they think. She goes on to explain that 'talking to the women themselves produced alternative insights from "below" as to how they experienced the surveying profession, and what was "really" happening within it'. (Greed, 1991, p. 12). Her conversations with the women become the key empirical component of the research, as she said she realised that the male reality was the norm and already 'known'.

Greed's position was declared upfront along with her interest in the research (1991, p. 13). She said 'I am in a sense both researcher and researched' (1991, p. 14) and 'I am purposely aiming at subjective accuracy'. Her methodological approach went on to explain the false perceptions of the surface reality of surveying, and she adopted an ethnographic approach to 'getting at the truth' (Greed, 1991, p. 14). The data was primarily 'soft' comprising personal observations, anecdotes, examples and reportage, backed up with material from journals and other literature. Greed explained that she necessarily included some sensitive and negative issues, but that she was also aware of the complexity of the issues and did not assume that the 'problem was the men'. She wanted gender to be seen as key, but also to acknowledge that gender exists within other realities such as class, race and individual characteristics – supporting the feminist assertion of being sensitive to diversity rather than professing a single female perspective. Her findings also illustrated the continuing need for a feminist approach to research in the built environment, as she admitted in her conclusion, 'It appeared that many of the men still see the world of surveying and the inhabitants of the built environment as male' (p. 180).

Other examples readers may wish to follow up include:

Louise Ellison – *Surveying the Glass Ceiling : An Investigation of the Progress Made by Women into the Surveying Profession* (1999).
Pat Turrell and Sara Wilkinson – *Building a Culture* (2005).
Ann de Graft-Johnson, Sandra Manley and Clara Greed – *Why do Women Leave Architecture?* (2003).

Caroline Whitzman – The loneliness of the long distance runner: Long-term feminist planning initiatives in London, Melbourne, Montreal and Toronto (2007).

Helen Lingard, Anna Sublet – The Impact of job and organizational demands on marital relationships satisfaction and conflict among Australian civil Engineers (2002).

Conclusions

This chapter commenced by stating that a brief overview only was possible of this enormously dynamic, highly diversified and thoroughly challenging area of research practice. Feminist research is research conducted for, by, and about women and, significantly, it is research that attempts to bring about change. An explanation of the terms and the sources of feminist research are outlined for the reader with an explanation of its relevance. The chapter showed that there are lessons to be learned for other traditional approaches especially with regards to ethics and objectivity. Approaches such as standpoint research, oral histories and diaries and participatory action research are important in this paradigm and are used to gain richer, deeper knowledge. It can be concluded that the built environment professions remain a significantly male-dominated sector and as such a gender focus on research is very much needed and will continue to be of utmost importance. Finally readers are referred to the works cited in the chapter for further readings, which will extend and deepen their knowledge base and understanding of this important paradigm.

References

Addelson, K. (1983) The man of professional wisdom, in Harding, S.G. and Hintikka, M.B.P. (eds) *Discovering Reality: Feminist Perspectives on Epistemology, Metaphysics, Methodology and Philosophy of Science*, Kluwer Academic, London.

Bell, D., Caplan, P. and Begum, K.W.J. (1993) *Gendered Fields : Women, Men, and Ethnography*, Routledge, London/New York.

Belenky, M.F., Clinchy, B.M., Goldberger, N.R. and Tarule, J.M. (1997) *Women's Ways of Knowing: The Development of Self, Voice and Mind*, Basic Books, USA.

Bordo, S. (1987) *The Flight to Objectivity: Essays on Cartesianism and Culture*, State University of New York Press, Albany.

Burns, D. and Walker, M. (2005) Feminist methodologies, in Somekh, B. and Lewin, C. (eds), *Research Methods in the Social Sciences*, Sage Publications, London.

Butler, J. (1990) *Gender Trouble*, Routledge, New York.

Carter, R. and Kirkup, G. (1990) *Women in Engineering: A Good Place To Be?*, Macmillan, Hampshire.

Chambers, R. (1993) *Rural Development: Putting the Last First*, Longman, London.

Code, L. (ed.) (2000) *Encyclopedia of Feminist Theories*, Routledge, London.

Cook, J. and Fonow, M.M. (1986) Knowledge and women's interests: Issues of epistemology and methodology, *Feminist Sociological Research, Sociological Inquiry*, 56(4), 2–29.

De Graft-Johnson, A., Manley, S. and Greed, C. (2003) *Why do Women Leave Architecture*, Research project funded by the RIBA and match funded by the University of the West of England.

Denzin, N.K. and Lincoln, Y.S. (eds) (2000) *The Handbook of Qualitative Research* (2nd Edition), Sage Publications, Inc., USA, ISBN 0-7619-1512-5.

Diamond, C. (1991) Knowing tornadoes and other things, *New Literary History*, 22, 1001–1015.

Duran, J. (1991) *Toward a Feminist Epistemology*, Rowman & Littlefield, Savage, MD.

Ellison, L. (1999) *Surveying the Glass Ceiling: An Investigation of the Progress Made By Women in the Surveying Profession*, Report on Research Funded by the Education Trust of the RICS, September 1999, RICS, London.

Freedman, E.B. (2002) *No Turning Back: The History of Feminism and the Future of Women*, Ballantine Books, New York, USA, available at http://noturningback.stanford.edu/quotes.html.

Gilligan, C. (1993) *In a Different Voice: Psychological Theory and Women's Development*, Harvard University Press, Cambridge, Massachusetts, and London, England.

Greaves, L., Wylie, A. and the Staff of the Battered Women's Advocacy Center (1995) Women and violence: Feminist practice and quantitative method, in *Changing Methods*, Burt and Code (eds), Broadview Press, Peterborough, Ontario.

Greed, C. (1991) *Surveying Sisters: Women in a Traditional Male Profession*, Routledge, London.

Greed, C. (1999) *The Changing Composition of the Construction Professions,* End Report of an ESRC pilot study, Occasional Paper 5, University of the West of England.

Hammersley, M. (1995) *The Politics of Social Research*, Sage, London.

Harding, S. (1986) *The Science Question in Feminism*, Cornell University Press, Ithaca.

Harding, S. (1991) *Whose Science? Whose Knowledge? Thinking from Women's Lives*, Open University Press, Milton Keynes.

Harding, S. (1993) Rethinking standpoint epistemology: 'What is strong objectivity?', in Alcoff, L. and Potter, E. (eds) *Feminist Epistemologies*, Routledge, New York.

Harding, S. (1998) *Is Science Multicultural?: Postcolonialisms, Feminisms, and Epistemologies*, Indiana University Press, Bloomington, IN.

Hartsock, N. (1983) *Money, Sex and Power: Towards a Feminist Historical Materialism*, Longman, New York.

Haslanger, S. (2000) Gender and race: (what) are they? (what) do we want them to be?, *Nous*, 34(1).

Hubbard, R. (1990) *The Politics of Women's Biology*, Rutgers University Press, New Brunswick, NJ.

Jaggar, A. (1989) Love and knowledge: Emotion in feminist epistemology, in Garry, A. and Pearsall, M. (eds) *Women, Knowledge and Reality*, Unwin Hyman, Boston.

Kalbfleisch, P. (ed.) (1995) *Gender, Power, and Communication in Human Relationships*, Erlbaum, Hillsdale, NJ.

Keller, E.F. (1986) How gender matters, or, why it's so hard to count past two, in Harding, J. (ed.) *Perspectives on Gender and Science*, Falmer Press, Lewes.

Kleiber, N. and Light, L. (1978) *Caring for Ourselves: An Alternative Structure for Health Care*, BC Public Health, Canada.

Lingard, H. and Sublet, A. (2002) The impact of job and organizational demands on marital or relationship satisfaction and conflict among Australian civil engineers, *Construction Management and Economics*, 20(6), 507–521.

MacKinnon, C. (1999) *Toward a Feminist Theory of the State*, Harvard University Press, Cambridge, Mass.

Millen, D. (1997) Some methodological and epistemological issues raised by doing feminist research on non-feminist women, *Sociological Research Online*, 2(3), available at: http://www.socresonline.org.uk/socresonline/2/3/3.html.

Morton, P. (2006) Culture and the student experience of women in Built Environment Higher Education, paper presented at *Gender and Science and Technology 12*, Brighton University, September 3–8, 2006.

Nelson, L.H. (1990) *Who Knows: From Quine to a Feminist Empiricism*, Temple University Press, Philadelphia, PA.

Reinharz, S. (1992) *Feminist Methods in Social Science Research*, Oxford University Press, New York and Oxford.

Rooney, P. (1991) Gendered reason: Sex metaphor and conceptions of reason, *Hypatia*, 6, 77–103.

Stanley, L. and Wise, S. (1993) *Breaking Out Again: Feminist Ontology and Epistemology*, Routledge, London.

Turrell, P.M. and Wilkinson, S.J. (2005) *Building a Culture*, RICS COBRA Conference, Queensland University of Technology, Brisbane, Australia, July 2005, pp. 1192–1202.

Turrell, P.M., Wilkinson, S.J., Astle, V. and Yeo, S. (2002) A gender for change: The future for women in surveying, paper presented at *FIG XX11 International Congress*, Washington, DC, USA, April 19–26, 2002.

Whitzman, C. (2007) The loneliness of the long distance runner: Long term feminist planning initiatives in London, Melbourne, Montreal and Toronto, *Planning Theory and Practice*, 8(2), 205–227.

Young, I.M. (1990) *Throwing Like a Girl and Other Essays, in Feminist Political Theory*, Indiana University Press, Bloomington.

Chapter Five
Approaches to economic modelling and analysis

Les Ruddock

Introduction

For researchers of the built environment, emphasis is normally on empirical research rather than theoretical research. This is not meant to demean the role of theory in research but to acknowledge the connection between theory and investigation that makes use of data or information. The work of such researchers may require some application of theory, which has been encountered in studies of economics, to a real-world problem in their field of study in the built environment.

Research work may involve a need to relate the construction industry to the general economic environment and to consider issues of construction and its relationship with the rest of the economy. As many academic programmes in the built environment have recognised the need for the inclusion of a study of fundamental economic concepts and theory, so there have evolved a number of introductory books on construction industry economics. A feature of such texts is a basic recognition that the construction industry is unlike the 'generalised' industry of economic theory. Many aspects of the industry do not fit into classical economic theory, especially at the microeconomic level. This has led to various attempts to develop concepts and consider alternative approaches to issues such as production theory appropriate to the industry – work in the area of 'lean production' being one such example.

The focus of this chapter, however, is a theoretical approach to the construction industry based on macroeconomic (and mesoeconomic) theory relevant to the construction industry. Researchers without any economic background, who wish to delve more deeply into basic macroeconomics should broaden their understanding by reference to a textbook such as Samuelson and Nordhaus (2004) or Sayre and Morris (2006).

Research into the economic processes involved in the field of construction economics often means applied research in the field to test the validity of hypotheses and this requires meaningful analysis of the data surrounding construction. A feature of this chapter is consideration of how the data are used to analyse the relationship between the construction industry and the wider economy.

General economic models

An economic model basically represents a scaled-down version of the big picture in order to understand important relationships between variables. The basis for macroeconomic models can be traced back to Leon Walras, a nineteenth-century French economist. Walras was adamant that one could not explain anything in an economy until one had explained everything. Each market (for goods, labour and capital) was connected to every other, however remotely. As an example, this interdependence is apparent whenever the effects of rising interest rates on the housing market are considered. Large, aggregate models may be developed, which span all the markets in an economy.

Leontief was one of the first to do more than just theorise about the complicated nature of this interdependence. In *The Structure of the American Economy* (Leontief, 1941) there was an appended table, in which were shown the flows of commodities and services between America's national industries, households and trading partners. In Leontief's blueprint, each industry was represented by an equation, with the inputs to an industry entered on one side of the equation and the industry's output appearing on the other. Since the output of one industry (steel, for example) serves as an input for another (construction), any equation cannot be solved without solving them all simultaneously.

Economists have put a lot of effort into determining the historical relationships between macroeconomic variables (such as inflation and unemployment) but the use of quantitative models has required academic economists to build for their models, foundations that would not shift when policies changed. To underpin such quantitative models, the foundations were to be found in the 'microfoundations' of macroeconomic behaviour. Everything that happens at the level of the economy as a whole is simply the sum of the actions of individual households or firms. If you know how the 'representative' firm or household makes its choices, you can forecast how the economy might respond to changes, even if that type of change had never happened before.

Relationships between economic variables – econometrics

It is common practice for economists to explain a theory in terms of an equation or set of equations. Modellers need to be explicit about the theoretical principles that underlie their simulations. But to compute an economic model, this theory has to be given concrete form, which often means that it must be spelt out in definite algebraic terms. Even basic economics textbooks present relationships between economic variables in an algebraic form. A natural consequence of this is to give quantitative values to these relationships. It may be noted that Alfred Marshall, one of the fathers of neo-classical economics, distrusted mathematics for this very reason. To be expressed in mathematical form, he complained, many important economic considerations had to be 'clipped and pruned until they resembled the conventional birds and animals of decorative art'. The most widely used tool of economists to determine empirical forms of theoretical constructs is that of econometrics. Theoretical economics may suggest that there is a relationship among two or more variables but applied economics demands evidence that the relationship is a real one observed in quantification of the relationship between the variables.

Econometrics simply means 'measurement in economics'. According to Charemza and Deadman (1997), the likely originator of the term 'econometrics' defined it as '… the unification of economic theory, statistics and mathematics …' (Frisch, 1936, p. 95). In practice, it includes all those statistical and mathematical techniques used in the analysis of economic data. The main target of using these statistical and mathematical tools with economic data is to attempt to prove or disprove certain economic propositions or to test or develop models.

The objectives of econometrics have been classically described by Christ (1966, p. 4) as: 'the production of quantitative economic statements that either explain the behaviour of variables we have already seen, or forecast (i.e. predict) behaviour that we have not yet seen, or both'.

In order to undertake econometric analysis, he determined the following to be prerequisites:

(1) Economic theory.
(2) Statistical data.
(3) A method that allows for the expression of the economic theory using the statistical data.
(4) A methodology, which tells us how to apply the estimation theory to the statistical data and how to decide whether this application has been successful.

On the basis of a model determined through this process, there are six properties that could be considered desirable in an estimated model:

Relevance, simplicity, theoretical plausibility, explanatory ability, accuracy of coefficients and forecasting ability. A 'good' model will display all of these properties to some degree.

In Chapter 13, various statistical concepts are considered, but it should be noted that any detailed explanation of statistical or econometric techniques is beyond the scope of this chapter and readers wishing to obtain a basic understanding of such techniques should refer to standard econometrics texts, such as Dougherty (2006) or Gujarati (2002). For those already in possession of some knowledge of basic techniques, Favero (2001) provides discussion and the practical illustration of techniques used in applied econometrics, and Charemza and Deadman (1997) offer non-specialised econometricians an intuitive understanding of recent developments in modelling.

Testing a model

Applied econometric work in practice should have, as a starting point, a model or an economic theory. From this theory, the researcher can develop an econometric model that can be used in an empirically testable form. Then, the next task is to collect data that can be used to perform the test and after that to proceed with the estimation of the model. After the estimation of the model is done, the researcher has to perform specification tests to make sure that the model used was the appropriate one, as well as some diagnostic checking in order to check the performance and accuracy of the estimation procedure. If those tests suggest that the model is adequate, then the next test is to apply hypothesis testing in order to test the validity of the theoretical predictions.

Figure 5.1 illustrates this procedure.

Choice of approaches

Until the mid-1970s there was no dominant stream of econometric methodology to guide research. Since 1975, there have been many attempts by a number of econometricians to build methodologies for econometric analysis. 'Implicit in these actions has been the notion that works along the prescribed lines would "better" econometrics in at least three ways. First, the methodology would (and should) provide a set of principles to guide work in all its facets. Second, by codifying this body of knowledge it should greatly facilitate the transmission of such knowledge. Finally, a style of reporting should naturally arise from the methodology that is informative, succinct and readily understood' (Pagan, 1987).

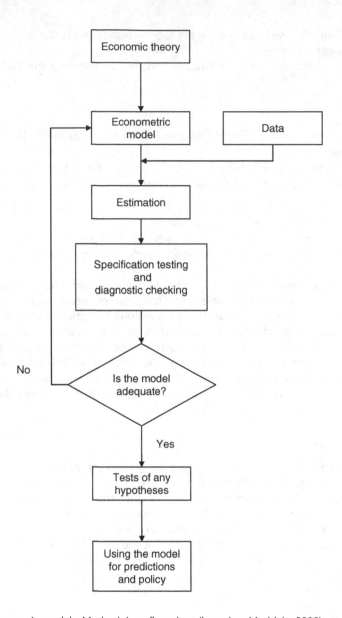

Figure 5.1 Economic models: Methodology flow chart (based on Maddala, 2000).

The methodologies, which have emerged over recent decades have attempted to design practical steps that should be followed in applied economic studies. These approaches have emphasised the relationship between economic theory and the statistical modelling of data. There are two main types of approach. First, the traditional approach which is 'bottom-up'.

This approach arose in the late 1940s from the work of the Cowles Commission (concerned with applied macroeconometric modelling) and was particularly concerned with the mapping between theory and data. The main notion behind this approach is that 'the model can never be a completely accurate description of reality; to describe reality one may have to develop such a complex model that it will be of little practical use. Some amount of abstraction or simplification is inevitable in any model building' (Gujarati, 2002, p. 454). According to this view, the model should be kept as simple as possible based on the notion that it should follow the underlying economic theory. After the estimation process, such a model is subjected to three levels of evaluation in order to determine whether or not it is satisfactory. A satisfactory model in this sense would be one which: (a) has coefficient signs, which correspond with the theoretical predictions (economic evaluation), (b) have both significant and good-fit coefficients (statistical evaluation), and (c) have residuals that do not suffer from autocorrelation or heteroscedasticity (econometric evaluation). If at least one of these points is violated, then the researcher should attempt to find another estimation technique that would improve the model. The second procedure in this case is to re-examine the model to ascertain whether important variables have been omitted from the model, or whether redundant variables have been included in the model, or to examine alternative forms of functional forms, and so on (Asteriou and Hall, 2007).

The second approach is popularly known as the 'top-down' or general to specific. This approach is closely related to a wider range of work on integrated and co-integrated systems. The methodology starts with a model with several regressors and then whittles it down to a model containing only the 'important' variables. The pivotal point of this methodology is the notion of data reduction, which is centred on the analysis of 'exogeneity'. In practical terms, the theory of general-to-specific attempts to explain how the data can be characterised in a way that is partial, or simpler than the true data-generating process without loss of information relative to the question of interest. This approach involves starting with as broad a general specification as possible and trying to explain how econometric models are derived from the data-generating process. The important question is: How does the researcher know what the final simplified model should be? The answer is that the simplified model should: (1) be data admissible; (2) be consistent with the theory; (3) use regressors that are not correlated; (4) exhibit parameter constancy; (5) exhibit purely random data (white noise); and (6) be encompassing (include all possible rival models) (Asteriou and Hall, 2007).

Data issues

Much of the early empirical work in econometrics was concerned, at the macroeconomic level, with business and economic cycles. The availability of data has, of course, always been a constraint on the application. It was the development of long runs of macroeconomic data that allowed new opportunities for econometric analysis of macroeconomic series.

Choice of statistical technique

The history of macroeconomic modelling started with the Dutch economist Jan Tinbergen. He showed how one could build a system of equations into an econometric model of the business cycle, using economic theory to derive behaviourally motivated dynamic equations and statistical methods to test the original data (Tinbergen, 1937). Statistics entered this discipline of economics and econometrics with the contribution of Tygve Haavelmo. He recognised, in the context of an economic model, that the joint distribution of all observable variables for the whole sample period provides the most general framework for statistical inference (Haavelmo, 1944). The most familiar form of macroeconomic evidence is statistical – aggregate data such as expenditure, output, employment statistics and so on. Once it has been established what data to use, there arises the question of how to analyse the data and how to confront it with the model. That is, is it better to find a model that fits the data or to use the data to reject models that are unsatisfactory? Whichever choice is made, the biggest question is often which statistical techniques to employ in order to estimate parameters and to decide whether the fit between the model and the data is satisfactory.

Such issues arise in any discipline, where statistical evidence is used but macroeconomics faces particular problems because of its distinctive characteristics – data are typically aggregative. This can take the form of aggregation over commodities or aggregation over time. Time-series data pose specific statistical problems, with a range of competing methods having been developed to solve them, whereas cross-section studies raise a different set of issues. Neither case conforms to the type of experimental situation for which classical statistical methods were developed.

Concerning data manipulation, the humble spreadsheet is still the most accessible means to study and analyse data. However, off-the-shelf spreadsheets have their limitations, lacking many of the functions required in modern econometrics, even though they can be used to run macros of unlimited complexity. Consequently, specialist software packages, such as *Stata*, for cross-section analysis, and *EViews* for time series are often needed for any analysis beyond the basic level.

Approaches and applications in the construction sector

To analyse economic relationships, different levels of approach can be considered. The distinction between the different 'levels' may appear arbitrary but, nevertheless, it is worth making a distinction between the development of (a) an *economic framework*, which permits the investigation of relationships without a full-blown, statistically rigorous analysis, and the formulation of (b) an *economic model* which may accompany the development of an econometric analysis.

An example of each of these two approaches is considered in the following sections.

(a) An economic framework: The Construction Sector (Mesoeconomic) approach: Under the auspices of the 'International Council for Research and Innovation in Building and Construction' (CIB), a Project Group was set up by to undertake the development of a framework for the analysis of the entire construction and property sector of the economy. The proposal for the Group was the development and testing of a mesoeconomic framework to supplement existing methods of viewing the sector. A framework was developed to consider a fundamental question and justification for the framework was based on the proposal of four hypotheses made about the construction sector. These hypotheses were rooted in economic theory. The validation of the framework was then made on the basis of populating the framework with data from nine countries.

The mesoeconomic method consists of using a unifying concept (the 'economic sector system'), to study the system implemented, and in applying a mesoeconomic method of analysis based on the following notions: the aim of construction activity, determining characteristics, groups of activities, profit formation, industry structure, operational configurations of players and institutional regulations (Carassus et al., 2006).

In its application in this specific context, the basic features distinguishing the mesoeconomic approach from the macroeconomic are summarised in Table 5.1.

	Areas and topics	Elements of analysis
Mesoeconomics	Industries	Theories of economic structure and change
		Environmental economics
	Groups	Theories of groups and associations
		Economic theory and politics
Macroeconomics		National economic accounts
		Economic stability and growth
	Total economy	Monetary theory
		International trade
		Macroeconomic distribution theory

Table 5.1 Scope of the mesoeconomic and macroeconomic approaches.

In terms of considering the overall aim of construction activity, the object of economic interest cannot simply be considered to be the 'act of building'. Rather, the issues involved in construction are more wide-ranging and represent a considerable economic and social challenge. It is a question of producing and managing the whole living and working environment. The entire built environment, as distinct from the natural environment, falls into the field of activity of construction. The meso-economic approach considers that building *per se* is not the only activity in this field. There is also management, maintenance, improvement, demolition, reconstruction, and so forth concerning the whole built environment.

The proposed hypotheses

A first hypothesis made about construction was that its principal aim is not to produce and manage necessary structures for people's living and working environment, but rather to produce and manage the services rendered to end users by these structures throughout their physical life-cycle (i.e. production, use, maintenance through to demolition). A review of the literature (such as Ive and Gruneberg, 2000; Hillebrandt, 2000; Manser, 1994), indicates that there may be various 'shaping' characteristics – the physical nature of the product, the structure of industry, the factors determining demand and price, the demand for single structures, the diversity of clients, the importance of maintenance, the geographical dispersion of worksites, the derived nature of demand, the long life expectancy of structures and the heterogeneity of production and techniques.

From these 'shaping characteristics', two are particularly important. A second hypothesis is that two major characteristics are: the fact that the demand for building products presents an extraordinary diversity and heterogeneity, and production is localised and static on site. Other elements characterising construction tend to be determined by these two structuring particularities.

A third hypothesis about construction is that, especially in the more developed countries, the stock of existing buildings, its optimisation and renewal, have become a central issue of construction activity. A significant indication of this evolution is the major role played by improvement and maintenance work in construction activity. A fourth hypothesis is that fragmentation is determined in particular by three factors: fragmentation of the demand, the degree of technical complexity and the capital intensity of the activity. Owing to these three criteria, any segment of the construction sector will be more or less fragmented into a large number of companies. This is thus a differentiated fragmentation. One of the forms this fragmentation may take is the eventual existence of construction sub-systems determined by the nature of the structures, their complexity, the type of market and the size of the construction firms.

Figure 5.2 shows the full life-cycle, including extraction, briefing, design, works and sale for varying types of projects and shows how these areas are interlinked with one another.

Comparison of the construction industry and construction sector approaches

The rationale for this mesoeconomic approach is based on the fact that the role played by the construction and property sector within the economy of developed countries has changed significantly over recent decades. It is no longer focused on large-scale production but on the services provided by the built environment. The requirements of sustainable development, which focus on the need to increasingly master medium and long-term consequences, not only regarding production, but also management of the works during their whole life-cycle have strengthened this change of role within the economy. This focus on the service rendered by the works calls for a new approach for the analysis of the construction and property sector.

Economic analysis has to take into account such recent evolution and all the participants involved in the life-cycle of building structures (not only procurement, design and production but also operation, maintenance, refurbishment and demolition). It is common, in construction industry analysis, to deal only with construction firms. Some research analysis has expanded the area of research to the materials industry but the service aspects and stock management firms have tended to be outside the scope of industry studies.

An overview of previous studies

The study of the construction industry and its role in the national economy has been extensively researched. At the macro level, existing assumptions persist that structural changes will emerge in the construction industry of a particular country as the national economy develops over time. Turin (1973), in his analysis of the role of the construction sector in economic development, presented a development pattern of the construction industry based on stages in an economy's development. The main aspects of the development pattern were that, in the early stages of development, the share of construction in national output first grows at an increasing rate and then at a decreasing rate with the level of national income. This 'S' shape pattern contrasts with Bon's (1992) inverse 'U' shape pattern in which the share of construction in national output increases in the early stages of development but ultimately will decrease in absolute and relative terms in more advanced industrial countries. Another important aspect of the development pattern derived from the latter work is that, whilst the share allotted to improvement and maintenance in total construction increases, the proportion for new construction decreases in the latest stages of development. Ruddock (2000, and with Lopes, 2006), using data collected from a large sample of countries representing all stages of economic development corroborated this proposition.

At the sectoral level, writers often analyse the way the construction firm operates within the sector's specific environment (see particularly, Gruneberg and Ive, 2000).

The construction sector framework approach can also be described as a 'cluster approach', as is illustrated by the fact that Figure 5.2 describes the main functions and regulations of a built environment cluster. Asset, property, facilities and transaction management is undertaken by services firms involved in ownership, operation,

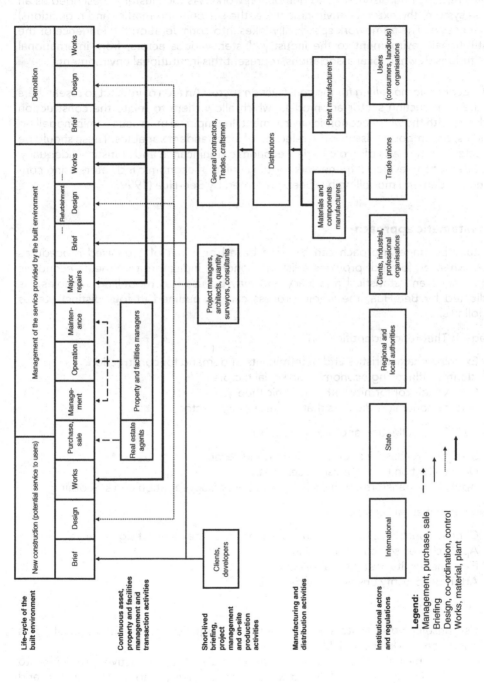

Figure 5.2 Construction sector system: Main functions and regulations (Based on Carassus et al., 2006).

maintenance, purchase and sales activities. Project management and on-site production are provided by services firms dealing with briefing, design, technical studies, co-ordination and control, and by construction firms involved in new construction, major repairs, refurbishment and demolition works. As the cluster is described as an open system, the external environment (i.e. the surrounding market and regulations) influences it. The framework specifically takes into consideration the influence of the institutional environment to the industry cluster. Various actors, from international institutions down to local associations, represent this institutional environment.

(b) *Economic modelling for the construction sector*: An area of interest to researchers of the construction sector are models, which allow them to relate the construction industry with the wider economic environment. In empirical macroeconomic modelling studies, one important issue is the selection of independent variables. These should be selected on the basis of two criteria – economic significance and statistical adequacy. An attempt to develop a framework for the choice of economic indicators using construction demand modelling has been presented by Bee-Hua (1999).

A systematic approach

A stage-by-stage approach can be used based on a set of organised procedures that, when performed, provides a list of economic indicators that meets economic significance and statistical adequacy and may be used as modelling variables. As indicated by Bee-Hua, the whole process can be outlined in four distinct stages as follows:

Stage 1: Theoretical identification

- Examine characteristics and determinants of demand for construction.
- Identify influencing economic and social factors.
- Check their conformity with economic theory.
- Select economic indicators that represent each factor.

Stage 2: Data collection and pre-processing

- Statistical re-classification of economic indicators.
- Data collection from published services.
- Apply data transformation such as, deflation, disaggregation and smoothing.

Stage 3: Statistical selection

- Choose variable selection methods and appropriate levels of significance.
- Apply chosen selection procedures.
- Examine results of statistical selection.
- Attempt to justify unselected indicators.

Stage 4: Usage

- Use unselected indicators whose statistical insignificance cannot be justified as additional modelling variables.
- Use variables that have been selected by the less restrictive procedure to compile an exhaustive list of indicators that satisfies economic significance and statistical adequacy.
- Use variables that have been selected by the more restrictive procedure to model demand for construction.

General characteristics from theory	Economic and social factors	Economic indicators
Cyclical fluctuations in the demand for new housing	Economic growth	Real GDP
		GDP per capita
		Productivity
Housing is a capital consumption good largely financed by long-term borrowing	Cost of borrowing	Interest rate
	Family formation	Household formation
	Population size	Population
	Property price	Property price index
	Level of income	Disposable income
		GDP per capita
		Wages and earnings
	Level of unemployment	Unemployment
	Existing housing stock	Housing stock (additions)
	Government intervention	Planning approval
	Rate of inflation	Consumer price index
Housing need is a basic determinant of demand	Construction price	Building tender price index
		Building material price index
	Mortgage credit availability	Money supply
		National savings
		Bank lending

Table 5.2 Determination of potential economic indicators (Residential demand).

Using influential factors as indicators, and taking, as an example, residential demand, Table 5.2 shows the designation of potential economic indicators derived from underpinning general characteristics based on theory.

The theoretical construct for the general characteristics underlying the demand for housing is based on consideration of both the basic determinants of housing demand from a theoretical approach (such as that found in Hillebrandt, 2000) or from observation of the specific market. The listed economic indicators for modelling relate to the general characteristics, enabling the determined quantitative relationships to reflect the supporting characteristics.

Conclusions

This chapter represents an attempt to explain some of the issues involved in undertaking economic analysis of, and attempting to model, the construction sector. Different approaches to the level of depth of analysis have been examined and two examples of such approaches considered.

Problems and issues facing the analyst of the construction sector are those common to researchers in other sectors, albeit to different degrees:

- *Non-availability of data.* The construction sector is particularly problematical in the availability of valid and reliable data – an issue examined in Ruddock (2000, 2002).

Identification and selection of appropriate economic variables or indicators.

- *Knowledge of economic theory or, in the case of the construction sector, a lack of appropriate theory.* An exploration of structural relationships may allow modellers to build more realistic models. Failure of models can be ascribed to any of these concerns – data availability, model specification or the fact that they have a weak theoretical base. For a model to be effective for practical purposes of forecasting and policy-making, any model needs to represent the data used and also represent the theory on which it is based.

Finally, a couple of questions may be asked by the modeller:

(1)   What might be the expectations of the users of a model?

Users of a model need not grasp exactly how a model works – any more than a car driver needs to know how a car engine works.
   Scepticism of econometric models needs to be overcome:

the ad hoc approach of many practicing econometricians to the problem of hypothesis testing and inference is illustrated by the popular image of much econometrics as a high R^2 in search of a theory. Garbage in garbage out is how many describe their own activity   (Desai, 1976, p. vii).

(2)   Can a theory be assessed purely from its explanatory power?

The assumption that the future will be like the past may, in normal circumstances, provide a reasonably good means of forecasting. But it is obviously useless as a means of understanding either the effects of policy or of macroeconomic shocks on the economy.

References

Asteriou, D. and Hall, S.G. (2007) *Applied Econometrics*, Palgrave Macmillan, Houndsmill.

Bee-Hua, G. (1999) Construction demand modelling: A systematic approach to using economic indicators, *RICS Research Paper Series*, 3(4), RICS, London.

Bon, R. (1992) The future of international construction: Secular patterns of growth and decline, *Habitat International*, 16(3), 119–128.

Carassus, J., Andersson, N., Kaklauskas, A., Lopes, J., Manseau, A., Ruddock, L. and deValence, G. (2006) Moving from production to services: A built environment cluster framework, *International Journal of Strategic Property Management*, 10(3), 169–184.

Charemza, W.W. and Deadman, D.F. (1997) *New Directions in Econometric Practice*, Edward Elgar Publishing Ltd, Cheltenham.

Christ, C.F. (1966) *Econometric Models and Methods*, J. Wiley and Sons Inc., New York.

Desai, M. (1976) *Applied Econometrics*, Philip Allan, Oxford.

Dougherty, C. (2006) *Introduction to Econometrics* (3rd Edition), Oxford University Press, Oxford.

Favero, C.A. (2001) *Applied Econometrics*, Oxford University Press, Oxford.

Frisch, R. (1936) Note on the term 'econometrics', *Econometrica*, 4, 95.

Gruneberg, S.L. and Ive, G.J. (2000) *The Economics of the Modern Construction Firm*, Macmillan Press Ltd, Houndsmill.

Gujarati, D.N. (2002) *Basic Econometrics* (4th Edition), McGraw-Hill, New York.

Haavelmo, T. (1944) The probability approach in econometrics, *Econometrica*, 12, 1–118.

Hillebrandt, P.M. (2000) *Economic Theory and the Construction Industry* (3rd Edition), Macmillan Press Ltd, Houndsmill.

Ive, G.J. and Gruneberg, S.L. (2000) *The Economics of the Modern Construction Sector*, Macmillan Press Ltd, Houndsmill.

Leontief, W. (1941) *The Structure of the American Economy 1919–1939*, MIT Press, Cambridge, Mass.

Maddala, G.S. (2000) *Introduction to Economics* (4th Edition), Wiley, New York.

Manser, J.E. (1994) *Economics: A Foundation Course for the Built Environment*, E & FN Spon, London.

Pagan, A. (1987) Three econometric methodologies: A critical appraisal, *Journal of Economic Surveys*, 1, 3–24.

Ruddock, L. (2000) An international survey of macroeconomic and market information on the construction sector: Issues of availability and reliability, *RICS Research Papers*, 3(11), 1–17.

Ruddock, L. (2002) Measuring the global construction industry: Improving the quality of data, *Construction, Management and Economics*, 20, 553–556.

Ruddock, L. and Lopes, J. (2006) The construction sector and economic development: The 'Bon' curve, *Construction Management and Economics*, 24, 717–723.

Samuelson, P. and Nordhaus, N. (2004) *Macroeconomics* (18th Edition), McGraw-Hill Education, Maidenhead.

Sayre, J.E. and Morris, A.J. (2006) *Principles of Macroeconomics* (5th Edition), McGraw-Hill Ryerson, Toronto.

Tinbergen, J. (1937) *An Econometric Approach to Business Cycle Problems*. Hemann & Cie, Paris.

Turin, D.A. (1973) *The Construction Industry: Its Economic Significance and Its Role in Development*, UCERG, London.

Chapter Six

Epistemology

Andrew Knight and Neil Turnbull

Introduction

For any academic researcher, questions surrounding the status of knowledge are a serious concern. Over thousands of years, various theories have been developed by philosophers to address these difficult questions, many of which still influence the way we undertake research today. The aim of this chapter is to introduce the reader to some of the key ideas and thinkers in epistemology.

So often, when postgraduate researchers start their journey, they feel lost in a sea of terms. For example, they may hear that other students are 'doing positivist research', or using a 'postmodern perspective', or hear students claiming that they still need to 'come up with a few hundred words on epistemology and ontology'. Obviously, this can be very daunting for a student who may have never studied philosophy before, particularly when terms are confusing and used inconsistently. Two common reactions to these new challenges can cause difficulties. First, the student may believe epistemology has no relevance to their project. We believe this is a mistake. Any student undertaking a research degree will normally need to convincingly argue that their thesis contributes to knowledge in a particular field. If the student has no idea what knowledge is, or how it is acquired in the context of the project, it is unlikely that the candidate will be able to produce a strong defence.

A second common reaction is to become so absorbed by the philosophy of research (an interesting subject) that a student loses his or her way, and the project fails to practically progress. So, it is important that students undertaking built environment research explore the epistemological assumptions underpinning research without getting completely bogged-down in irresolvable philosophical problems. Many universities now have excellent research training courses, which often include sessions on research philosophy. These forums also provide excellent opportunities for testing and exploring your ideas with other students, who often come from different disciplines. However, many of the textbooks covering epistemology are aimed at philosophy students and are, therefore, quite difficult to access for non-specialists. Hence, the aim of this chapter is to act as a bridge for built environment researchers interested in epistemology.

This chapter starts with an examination of the key terms. However, to provide readers with a long list of words associated with epistemology is unlikely to be useful for those attempting to develop a deeper understanding. Terms such as 'empiricism' and 'rationalism' need to be understood in context. Hence, in this chapter we have aimed to provide an historical roadmap. In this way, the terms currently used in debates concerning methodological issues can be seen to have developed through various key thinkers. This historical approach also helps readers interested in certain areas to identify key thinkers for further study. Three major eras are considered: classical, modern and postmodern. The chapter concludes by arguing that since built environment researchers occupy an applied field of enquiry, rather than a formal academic discipline, a lack of clarity surrounding epistemology can result in the application of inappropriate quality criteria and audience misunderstanding.

Concepts

The term *epistemology* is derived from the Ancient Greek words *episteme*, which means *knowledge*, and *logos*, which can be approximated to the word *account*. As a sub-discipline of modern philosophy, epistemology is principally concerned with theories of knowledge. These theories attempt to answer questions surrounding the nature of knowledge, its limits and how we acquire it. In our everyday lives we constantly claim to know things. This may be as trivial as claiming to know there is no milk in the refrigerator, or as important as claiming to know that the bridge you have designed will not fail under normal circumstances. Or more fundamentally, it may be that you are claiming that God does or does not exist. So, in addition to knowledge claims made at the completion of specific research projects, where we may be claiming to contribute to knowledge, we are all, to some degree, epistemic agents in our everyday lives. This then leads to the important question: what is knowledge?

In philosophy, knowledge is typically defined as a 'justified true belief'. For instance, you may believe that interest rates will fall next month. If, in one month's time, interest rates do fall, is it fair to say you had knowledge that they would fall? The answer is: it depends on your justification. A true belief by itself does not constitute knowledge; a true belief requires an additional ingredient: justification. In this illustration your justification for your belief may have come from a variety of sources. For example, it may be that you had just spoken to all the members of the Bank of England Monetary Policy Committee that set interest rates. Alternatively, you may have randomly guessed the change in interest rates. In knowledge claims, it is therefore obvious that the source and quality of our justifications are very important, especially if we expect to convince other people. Since the Ancient Greeks, there has been disagreement amongst philosophers about how we should derive our knowledge and this difference in opinion can be used to broadly group various schools of thought.

One way of categorising epistemological theory is to compare whether the quality of the justification for a belief is the fundamental issue, against the condition or context in which beliefs are developed. The first of these is known as the normative tradition, which is the more dominant school, the second is termed the naturalistic tradition (Klein, 2000). However, within these categories there are subdivisions. For example, in the normative tradition a further distinction can be drawn between foundationalism and coherentism. In foundationalist epistemologies there is an assumption that knowledge rests on a basic, or foundational, belief. These unshakable basic beliefs act as a substructure for various other inferred non-basic beliefs. Coherentist epistemologies, on the other hand, hold that there are no foundational beliefs, only beliefs that support each other; the justification for the belief revolves around the match between it and other beliefs. Moreover, to take one branch of this categorisation further, it is possible to broadly split foundationalism between the rationalists and empiricists. Generally speaking, empiricists believe that knowledge is derived through the five senses; namely, sight, smell, hearing, touch and taste. It should be evident that this theory of knowledge closely fits the approach taken to knowledge acquisition in the experimental sciences. We often hear that there is no 'empirical support' for a claim. Rationalist approaches, however, assert that humans derive their basic beliefs not through the senses but from rational thought. This model of knowledge is used in mathematics, but many philosophers have argued that the rational capabilities of the mind are the best method for generating knowledge in a far wider field.

Many of the issues discussed above also raise what are known as ontological problems. Briefly, ontology is concerned with 'existence or being' and what we assume to exist clearly has implications for what we claim to know, and vice versa. These metaphysical questions on the nature of existence will be evident in the following sections outlining various philosophers' approaches to knowledge.

As may be evident from the above taxonomy, even labelling various groups of theories can be a complex task. Hence, we feel the most appropriate manner for the non-specialist to appreciate the challenges of epistemological issues in their own research is by first developing a broad overview of the chronology of thought in the area. As stated above, epistemology is a large and mature area of philosophy and in this chapter we can provide only the shortest of reviews. Nevertheless, by providing an overview, it is hoped that the reader will start to see the connections, or lineage, between some of the current methodological discussions in our own disciplines and the work of philosophers over the centuries. We feel that increasing awareness of the various traditions in epistemology can only serve to enlighten many of the questions being debated today, such as the validity of sample sizes, the appropriateness of qualitative methods, issues of generalisation and the status of non-empirical research to name just a few. Many of these issues are raised in other chapters in this book; the aim here is to provide a foundation for them.

Chronologically, epistemology can be classified into two orthodox schemes: classical and modern. Classical epistemology links questions of knowledge, especially the problem of who can be said to legitimately know, to more general problems of ethics and politics. Modern epistemology, in opposition, strives to be more value free and speak for and on behalf of the general modes of theoretical knowledge produced by the natural sciences, usually by striving to place such theoretical knowledge on secure incorrigible foundations. In what follows, we offer something of a Cook's Tour through the history of epistemology starting with Ancient Greeks and ending with recent criticisms directed against epistemology by the so-called *postmodern* philosophers of today. Although the term postmodern epistemology is often referred to in contemporary academic discourse, we take the view that postmodern analysis represents a radically anti-epistemological development. However, what should be clear is that the debates surrounding the justifications for our beliefs are as relevant today as they were 2500 years ago.

Classical epistemology

Classical epistemology can be divided into two main forms: Platonic and Pyrrhonic. Platonic epistemology, unsurprisingly, has its origins in the ideas of Plato (427–347 BC), or more accurately Socrates (469–399 BC) as represented in the works of Plato. Plato's epistemology drew upon a fundamental assumption of pre-Socratic Greek thought: that the intellectual powers of the human mind are of such magnitude that they could, given the appropriate training and education, apprehend the true nature of things and especially the true nature of ourselves. According to Plato/Socrates most people do not use their intellectual powers to good effect and as such cannot provide a true and justified account, or *logos*, of themselves and the world in which they live. They live, according to Plato, in a state of *akrasia* or self-delusion, which results in people acting against their own better judgement. According to Plato, it is the task of the philosopher to expose these illusions so that individuals can come to know themselves more authentically. That is, to become wise.

Thus, in the Platonic scheme the opposite and opponent of knowledge is general public opinion, or *doxa*. In fifth-century Socratic Athens, the idea that all knowledge is fundamentally *doxa* was peddled by groups of itinerant teachers known as the Sophists – the origin of the English word *sophistry* meaning superficially plausible, but poor reasoning. In Plato's view the sophists represented knowledge corrupted by power, money and rhetorical excess. True knowledge, for Plato, is not tied to or determined by the vagaries of public tastes, but is timeless, abstract and universal: that is, rational. According to Plato we know because we 'partake' in the universal essence of thing. Thus for Plato, true knowledge is never particular, never simply about this or that, but essential, general and fundamental. This conception lives on today in the epistemology of the natural sciences.

For Plato, to know something is to know its underlying unified form or general idea; and this, in his view, is something that lies behind and beyond the deceptive realm of appearance. For Plato, the senses, as they are fundamentally tied to this realm, produce poor candidates for knowledge and this is why Plato makes a sharp distinction between sensation 'aesthesis' and knowledge 'episteme'. Plato claims we cannot, in any ordinary way, experience forms, we can only intuit them, or as he puts it 'recollect them'; a process that he refers to as 'anamnesis'.

According to Plato, coming to know always involves processes of bringing to mind what, in a deep and inarticulate way, we somehow already know innately. Interestingly, Plato's epistemology carried with it significant political implications. Although this may be difficult for us to appreciate today, classical epistemology is a radically anti-democratic discipline in that it restricts knowledge to an elite cadre of thinkers. For Plato, democracy is the rule of the mob, of opinion, and therefore democracy is a mode of politics productive of error and illusion rather than knowledge *per se*. The tensions between power, politics and knowledge are still evident today in many of the standpoint epistemologies adopted by groups such as feminist researchers.

A contrasting approach to Platonism is Pyrrhonic epistemology, or perhaps better anti-epistemology. This tradition has its roots in the ideas of the so-called Sceptics, especially Phyrro (*c*. 360–270 BC) and Sextus Empiricus (*c*. AD 150–225), but is still evident today in the ideas of postmodernists. The Sceptics believed that the kind of general and universal knowledge sought after by the Platonists is impossible, and in general they believed that there is no way to resolve disputes between different ways of making judgements about the world. This is because, according to the sceptics, every argument for something carries with it an equal but opposite argument against it. Thus, the philosophical certainty associated with Platonism is fundamentally mistaken, and the only appropriate epistemological response is to suspend judgement (*epoché*). For the sceptic, nobody does, in fact, know anything because nobody can, in fact, know anything. As a result, those who understand the true nature of knowledge must develop the faculty of non-expression (*afasia*) in order to achieve a tranquillity of the soul that the sceptics termed *ataraxia*.

The idea that knowledge can and should have a value for practical living can also be discerned in Aristotle's work. Although in many ways Aristotle (384–322 BC) was Platonist, in that he gave priority to universal theoretical knowledge over and above practical knowledge, he recognised that knowledge is never purely and simply theoretical, but has an important practical aspect. Hence, Aristotle claims that it is important to distinguish between theoretical knowledge 'theoria' and practical knowledge 'praxis'. Moreover, for Aristotle there are three kinds of practical knowledge. First, there is knowledge required to make things 'techne'. Second, there is knowledge required to create or imagine things 'poesis' and finally, there is knowledge required in order live a good life 'phronesis'.

Again, for those of us with modern epistemological sensibilities the latter notion especially sounds rather odd, but this idea became an important motif in contemporary hermeneutics, especially the work of Gadamer (1979). He argued that with the advent of the massive capitalistic expansion of work and industrial production in the seventeenth century, 'techne' has come to dominate all other forms of practical knowing and as a result the other two types of knowledge have been largely marginalised and in some cases forgotten.

The above section has highlighted a few of the greatest thinkers of the classical era. It should be evident that many of the epistemological issues discussed by these philosophers still challenge us today. After the Ancient Greek civilisation, philosophy became an apologist for faith in the work of the mediaeval scholastics before the reawakening of the Renaissance in the fifteenth century, often seen as the birth of the modern world.

Modern epistemology

Rationalism

The ideas of the sceptics were rediscovered in the fifteenth century and enjoyed a place of revival in the sixteenth century. They formed a key aspect of the intellectual background that facilitated the emergence of modern epistemology. This is especially the case in the epistemology of those like Renée Descartes (1596–1650), who saw the ideas of the sceptics as a threat to the then developing scientific worldview. Other rationalists of the modern era included Leibniz (1646–1716) and Spinoza (1632–1677).

Descartes was aware that the scientific understanding of the world presupposed and required certain knowledge, and that this was, in his view, under threat as long as scepticism was deemed to be both plausible and possible. Descartes' strategy was to turn the sceptics' doubts about the possibility of knowing anything for certain against themselves, by bringing into play his so-called 'method of doubt'. According to Descartes there was a rational method for uncovering the forms of certain knowledge that science requires; a method premised on the idea that if any idea or claim could stand the most extreme of doubts then we can justifiably claim that we know that claim for certain. Descartes' method involved calling into question, 'rejecting as false', all those beliefs that are capable of being doubted, until we arrive at a belief that cannot be doubted on pain of contradiction or self-refutation. He thought that this final 'indubitable belief' could function as a foundation upon which other beliefs, derived from it by a process of logical deduction, could rest.

To assist him in this task, Descartes invented an imaginary interlocutor, the malignant demon, *malin genie*, whose task it was to make hypothetically false each and every belief that Descartes held to be true. Descartes famously argued that there is one belief that even an all-powerful malignant demon could not make false, and that is the *cogito* or the belief that 'I think'. This belief, in his view, is incapable of being doubted, because even to doubt that one is thinking is still in some way a mode of thinking. Therefore, Descartes argued that he was incapable of doubting that he is a thinking thing. Thus, Descartes states of his demon, 'let him deceive me as much as he can, he will never bring about that I am nothing so long that I think that I am something' (Descartes, 1996, p. 17).

Hence, for Descartes, rational certainty is the *foundation* of knowledge and the answer to the sceptic's global epistemological doubts. Descartes was thus a *foundationalist* philosopher because he thought that valid knowledge needed to rest on secure foundations, and he thought that knowledge required a first principle, an Archimedian point, upon which the whole edifice of knowledge could be derived and ultimately rest. The paradigmatic form of such foundational clear and distinct knowledge for Descartes was mathematical knowledge and the forms of proof with which they are associated. Mathematical, especially geometric notions, for Descartes possessed the requisite certainty required for the foundations of knowledge, and mathematical proofs were procedures that allowed for this certainty to be maintained in thought.

At this point it is useful to highlight the term *a priori* meaning prior to, or independent of, experience. The foundational, clear and distinct ideas in subjects such as mathematics can be known without going out to perform experiments. Hence, reason alone provides us with these *a priori* truths. This is in contrast to empirical knowledge, often termed *a posteriori*, meaning after experience. But this does not just apply to mathematics. For example, the statement 'all bachelors are unmarried men' is true by definition and hence *a priori*, but 'clients are, on average, more satisfied with projects completed under a traditional form of contract', is clearly impossible to know without data and is, hence, *a posteriori*.

Empiricism

Empiricism is the philosophy associated with the ideas of John Locke (1632–1704) and the Enlightenment philosophers George Berkeley (1685–1753) and David Hume (1711–1776). It is also associated with the ideas of the so-called positivists in the twentieth century. Generally, in empiricist accounts of epistemology any belief can count as knowledge if, and only if, it is grounded in sets of actual or possible *experiences*. More specifically, for most empiricists, in order to be justified in knowing that something is the case, one must be able to directly perceive that, that thing is the case. Seeing, hearing, tasting, smelling and touching are the routes to knowledge according to the empiricist. Empiricism is also a foundationalist epistemology in that it attempts to found knowledge on the certainties, or incorrigibility, associated with sense perception.

Although we think in abstract generalities, or concepts, we experience concrete specific things. Thus for the empiricist, unlike the classical Platonist, authentic knowledge begins with knowledge of concrete particulars, that is with an experience of this or that – this object, that person. The important point about our sense data is that it is direct sensation; it is not inferred from another source and so is an incorrigible foundation. This leaves the empiricist with a philosophical problem. As for most modern epistemologists, knowledge is only useful, valid and significant if it is linked to thought in some way – that is if it articulates the world at abstract and universal levels and from no particular point of view (see Nagel, 1986). Therefore, for empiricism to be a valid theory of knowledge it must show how we are to move or infer from concrete experience to abstract theoretical thought; that is from concrete to the universal knowledge.

For the early empiricists such as Francis Bacon (1561–1626) there is a method of inferring the universal from the particular; the method of induction. This method, crudely, states that after *n* – where *n* is a sufficiently large number – experiences that *a* is *b*, we are justified in claiming that all As are B. For example, if from a position of ignorance, I see thirty people from a particular culture behaving in a particular way, I may feel warranted in concluding that all people from that culture behave in that way.

Hume (1975) was the first to note that induction, as a cognitive process, could never warrant the assertion of general theoretical knowledge, as induction could not bridge the gap between the concrete experiential particular and the abstract universals of thought. For Hume, there was no rationality, no true preserving aspect, to inductive inferences; no such thing as 'inductive proof', and for him induction was merely an innate mechanism of the mind that generated *expectations* about the future relation between particular phenomena; expectations that we habitually treat as, and take for, universal truths. For Hume, all forms of empiricist epistemology that were based upon the idea that knowledge can be generated inductively lead to a scepticism about the human ability to apprehend general universal knowledge, and hence a scepticism about the status of knowledge itself. Hume was thus a sceptic, in that he denied the possibility of certain universal knowledge.

Hume's sceptical arguments unmasked many of empiricism's epistemological pretensions. The task for subsequent empiricists has been to articulate an epistemology founded in and justified by experience, but without attempting to derive the abstract and general levels of this epistemology by means of inductive reasoning. In the twentieth century, the philosopher Karl Popper (1959) attempted to circumvent this problem by inverting the inductivist conception of knowledge. He claimed that although experience did produce epistemically significant particulars, these discrete experiential particulars could not, logically, be induced into theoretical universals. For Popper, matters are the other way around; it is theoretical universals, as hypothesis, that have priority in both time and fact, and experiences only provide ways of testing for the truth value of these. Thus for Popper, the epistemological success of science is underpinned by a non-inductivist rationality; a rationality that he termed *hypothetico-deductivism*, which views the epistemological significance of empirical sense-experience negatively, that is as opportunities for putting hypotheses to the test.

For Popper, knowledge is always a form of guess-work or prediction. Good and valid general theoretical knowledge should be able to make concrete predictions about the world that can then be tested against experience, to the extent that the function of experience is to decide between competing predictions. Hence, experience can never justify knowledge of any general theoretical claim about the world, but it can refute knowledge claims. That is it can show that we do not know them because they turn out to be false. However, Popper's solution does not avoid the sceptical charge that Hume aimed at more traditional forms of empiricism. He still, like Hume, claims that experience can never provide grounds for general theoretical knowledge about the world; not, anyway, in the sense of providing positive grounds for its assertion.

Popper's account is related, although on close inspection it is more than slightly different to, what is often referred to as 'positivism'. Whereas Popper thought that sense experiences could only refute general theoretical claims about the world, the positivists believed that they could positively confirm them. Positivism owed a big debt to empiricism, in that it viewed all knowledge as tied to observational forms of 'verification', but instead of founding knowledge on sense experience, the positivists attempted to found knowledge on the methodologically ordered experiences associated with scientific experimentation. Hence, the positivist project was also epistemologically foundationalist in that it attempted to justify the epistemological significance of scientific theories by appealing to the incorrigible experiential certainties produced by scientific methods (Schlick, 1959). This is because for the positivists (see also, Neurath, 1959; Waismann, 1959), the experimental methods of science were the 'least poisoned well' from which the epistemic raw materials required for theory construction could be drawn. In the positivist scheme of epistemology, scientists essentially make predictions, that is, Schlick states:

> [s]cience makes prophesies that are tested by 'experience'. Its essential function consists in making predictions. It says for example: 'If at such and such a time you look through a telescope adjusted in such and such a manner you will see a point of light (a star) in coincidence with a black mark (cross wires).' … This means that we make an anticipated confirmation, we pronounce an expected judgement of observation.   (Schlick, 1959, pp. 121–122)

Postmodernism and the critique of epistemology

One of the problems with rationalist forms of foundationalism is the problem of historical and cultural variability. Consider again the Cartesian thesis that knowledge must be founded on the self-evidence of clear and distinct ideas. One of the problems with this idea is that what seems 'clear and distinct' at one point in history and in one location may seem opaque, confused or simply downright self-evidently false at another, and vice versa; that is, what seems a priori impossible at one historical juncture may seem necessary and desirable at another. One famous example comes from the work of another famous rationalist epistemologist Immanuel Kant (1724–1804). According to Kant, space is self-evidently Euclidean in that it is an infinite three-dimensional expanse, an expanse where the shortest distance between two points is a straight line and parallel lines never meet. However, in the twentieth century, and in the light of recent innovations and discoveries in relativity theory and quantum mechanics, it is now widely accepted that Euclid's ideas are only one theory of the nature of space and that other theories may be better, especially when we take into account some of Einstein's ideas on the nature of gravity and its ability to 'curve space'. Thus, many physicists prefer non-standard Riemannian geometries over and above those of Euclid; geometries where spaces are curved and parallel lines do meet at infinity. Moreover, quantum physics has also given rise to a certain loss of faith in scientific objectivity and certainty. As a result of quantum physics, scientific knowledge is increasingly viewed as simply one historical paradigm of knowledge, an idea made famous by the historicist philosopher of science Thomas Kuhn (1922–1996).

This shows that our time knowledge is no longer a static and timeless structure but is dynamic; ever-changing and ever-shifting. This is a result, in part, of the evidential and exploratory dynamism of experimental science. Moreover, as a myriad of technological changes have enabled people to experience the core ideas and ways of seeing the world that constitute other cultures, many have become aware of the geographical specificity of given ways of knowing, and that there may be other ways of knowing that do not fit into western rationalist or empiricist models and approaches. Suffice it to say, that many philosophers today are unimpressed with modern foundationalist epistemological projects. The so-called postmodern philosophers, for example Richard Rorty (1931–2007), have argued that there is no neutral and impartial standpoint from which to view the world because all perspectives on the world are mediated by language and culture. That is, knowledge is always relative to the interests at large in culture and so there is no such thing as timeless universal knowledge; only knowledge constructed, made or produced for specific purposes at particular times. This idea can take strong and weak forms, some arguing that there is no such thing as knowledge, only opinions, others that there is knowledge, but it always provisional and revisable in the light of new information.

Either way, the broad implication of this relativist critique of modern foundationalism has been broadly pragmatist in orientation. That is, for these critics knowledge cannot be judged according to rational criteria but only according to its general utility, or its cash value. In such accounts individuals and groups use a particular theory or concept, not because it was closer to the truth and hence a better candidate for knowledge, but because it allows you to do what you and fellow researchers, and what your epistemological community, wants you to achieve.

The postmodern critique of epistemology also emerges out of a series of reflections derived from the work of Friedrich Nietzsche (1844–1900). Nietzsche criticises those who want to link knowledge to timeless truth and transcendent objective realities, and he points out that once we recognise that truth has a history, or, as he states, a genealogy, then we can see knowledge as created rather than 'discovered'. Moreover, once the created nature of knowledge is recognised then the paragon of epistemo-logical virtue is no longer the disinterested self-reflexive thinker of Platonism and Cartesianism, but the artist. The basic idea of this school of postmodernism, an idea that we find in a good deal of contemporary post-structuralist thought, is that once we recognise the historicity of knowledge then knowledge loses its traditional philosoph-ical seriousness and becomes a matter of taste, something lucid or playful, rather than a matter of sober rationality. Knowledge becomes a matter of what Nietzsche termed 'the yes and no of the palate' rather than the outcome of considered rational reflection or deliberation.

For many, this kind of postmodern rejection of the very possibility of epistemology is extremely depressing in that it seems to threaten to undermine the very foundations of intellectual life. For how can we function intellectually without some idea that intellectual endeavours detect, discover or disclose new kinds of knowledge? However, although postmodernism seems to point towards a trite sophistry, which exclaims intellectual endeavours are futile because everything is a matter of opinion, it also helpfully suggests that knowledge is not separate from its historical contexts and the worlds in which people encounter things, others and themselves. It suggests, quite rightly, that knowledge is not a detached matrix of concepts and ideas, but is embedded in real histories and real lives and is thus an expression of the worlds in which people live. Thus the postmodernists show that once we pay attention to the history of knowledge, it is hard to separate knowledge from our surrounding environment (worldhood), and thus from ontology, as traditionally conceived.

Conclusion

The above sections illustrate that the subject of knowledge is complex and has a long history. Additionally, we have outlined different traditions in the theory of knowledge and some of the linkages between them. However, with such a vast area, this chapter can only map ideas briefly. In terms of broadening your understanding of general philosophy Turnbull (1999) and Osborne (1992) provide enjoyable and entertaining introductions. An excellent text aimed at final stage philosophy undergraduates is *Philosophy 1* (Grayling, 1995), which includes very good chapters on epistemology and methodology. For a very accessible book on epistemology try *Epistemology the Theory of Knowledge* (Cardinal et al., 2004). Good texts aimed at more advanced students, although still termed *introductory*, include Dancy (1985) and Audi (2002). Students should also explore the original works of philosophers; many texts such as Descartes' *Meditations* are widely available and very economically priced as paperbacks.

The problem for researchers in the field of the built environment is that their field of study covers a vast range of subjects and approaches. In this sense, the built environment is clearly not a discrete discipline with its own standard approaches to philosophy, methodology and methods. Extending Becher's geographical metaphor of academic disciplines (Becher, 1989), it is more like a large sprawling county, which encompasses parts of many cities and towns. For example, researchers in the built environment step into methods from mathematics, natural sciences, social sciences, arts and humanities. Of course, many research projects draw from more than one discipline causing further complications for those attempting to defend their ideas.

Hence, it is our argument that it is very important that researchers in these applied fields of enquiry, collectively termed the built environment, make their methodological and epistemological assumptions as clear as possible. This, of course, is particularly important for those defending a doctoral thesis. These decisions should not amount to a fashionable *pick and mix* of terms, but should be grounded in the genuine, and defendable, thoughts of the researcher and the subject of enquiry. It is also important that the whole methodological position forwarded by the researcher is coherent. For example, to argue that your research is based on anti-realist ontological assumptions and some form of post-modern theoretical position is likely to undermine a methodology based on questionnaire surveys including significance testing of Likert scales. It should be clear that in designing a methodology to investigate a problem, the researcher is building an edifice of assumptions around claims to knowledge and these assumptions should be explored and justified where appropriate.

One of the reasons it is critical to explore and justify claims to knowledge is that your work is always open to the judgements of others. All research is open to criticism. However, if you are attempting to persuade someone else of the quality of your research, it is necessary that the criteria on which you will be judged are accepted; if this is misunderstood by those reviewing your work, complications can arise. For example, this is often the case where research using a predominately qualitative methodology is judged against epistemic criteria developed for the judgement of quantitative work. Typically, questions around internal and external validity may be used to undermine the evidence of qualitative work if judged against traditional scientific epistemological criteria. To combat this, and to avoid charges of extreme relativism, Lincoln and Guba (1985, 2000) argue that alternative criteria are necessary for qualitative enquiry (but of course the postmodernists argue that there is no way of avoiding the charge of relativism), and that it is against these that quality judgements against individual studies should be compared. Example criteria include credibility, transferability, dependability and confirmability.

In some areas it will always be difficult to produce work in a vein that is contrary to the dominant approach. However, over time, as fields of enquiry are populated by researchers trained in a greater diversity of disciplines, acceptance of alternative assumptions and methodologies grows. For example, until a few years ago in the field of construction management research a majority of published research followed a positivist, or pseudo-positivist, approach. But, as the field matures, a greater range of methodologies and epistemologies are now becoming established

The key point to remember is that assumptions around existence and knowledge are clearly embedded in all forms of enquiry. If a researcher wants to have his or her evidence and arguments taken seriously, then it may be necessary to explore these assumptions to locate the work within an appropriate tradition. This should help to ensure the work is judged against the appropriate epistemic criteria when it is considered by the wider epistemological community.

References

Audi, R. (2002) *Epistemology: A Contemporary Introduction* (2nd Edition), Routledge, London.

Becher, T. (1989) *Academic Tribes and Territories: Intellectual Enquiry and the Cultures of Disciplines*, Oxford University Press, Buckingham.

Cardinal, D., Hayward, J. and Jones, G. (2004) *Epistemology the Theory of Knowledge*, Hodder Murray, London.

Dancy, J. (1985) *An Introduction to Contemporary Epistemology*, Blackwell, Oxford.

Descartes, R. (1996) in Cottingham, J. (ed.) *Meditations on First Philosophy: With Selection from the Objections and Replies*, Cambridge University Press, Cambridge.

Gadamer, H.-G. (1979) *Truth and Method*, Sheed and Ward, London.

Grayling, A.C. (ed.) (1995) *Philosophy 1: A Guide through the Subject*, Oxford University Press, Oxford.

Hume, D. (1975) *Treatise on Human Nature*, Clarendon, Oxford.

Klein, P.D. (2000) *Epistemology In Concise Routledge Encyclopaedia of Philosophy*, Routledge, London, pp. 246–249.

Lincoln, Y.S. and Guba, E.G. (1985) *Naturalistic Inquiry*, Sage, London.

Lincoln, Y.S. and Guba, E.G. (2000) Paradigmatic controversies, contradictions and emerging confluences, in Denzin, N.K. and Lincoln, Y.S. (eds) *Handbook of Qualitative Research* (2nd Edition), Sage, London, pp. 163–188.

Nagel, T. (1986) *The View From Nowhere*, Oxford University Press, Oxford.

Neurath, O. (1959) Protocol sentences, in Ayer, A.J. (ed.) *Logical Positivism*, Free Press, New York.

Osborne, R. (1992) *Philosophy for Beginners*, Zidane Press, London.

Popper, K. (1959) *Logic of Scientific Discovery*, Hutchinson, London.

Schlick, M. (1959) The foundation of knowledge, in Ayer, A.J. (ed.) *Logical Positivism*, Free Press, New York.

Turnbull, N. (1999) *Get a Grip on Philosophy*, Weidenfeld Nicolson Illustrated, London.

Waismann, F. (1959) How I see philosophy, in Ayer, A.J. (ed.) *Logical Positivism*, Free Press, New York.

Chapter Seven

Scientific theories

Göran Runeson and Martin Skitmore

Introduction

The aim of this chapter is to look briefly at theories in general, before concentrating on scientific theories – how they are used, structured, tested and verified. The research that uses the kind of scientific theories we are concentrating on is often referred to as quantitative research but for the sake of completeness, we will also discuss theories in the so-called qualitative research.

Theories are an absolutely essential part of our daily life. They help us to make sense of the enormous mass of information and perceptions we are bombarded with every day. Theories help us to recognise, identify and classify things and events, to understand, explain, relate and to make predictions. They give us context and hierarchy. In short, theories combine to make up our understanding of the world. We have theories for all purposes, theories that say that 'if you heat up a metal rod, it will expand' or 'the time required to make a decision is in inverse proportion to the money involved' or 'the earth is flat' or that 'if you sin, God will punish you'.

The philosophy behind theories

With so many different roles for theories, there is a corresponding array of different types of theory, based on different philosophies and different uses of the theories. The same applies to scientific theories. There is no single definition of 'Scientific Theories' beyond the general proposition that they are derived through scientific methods, but there are many methods based on many different, sometimes conflicting, philosophies and methodologies of science. What we will do here is to discuss the characteristics and uses of scientific theories to illustrate what we think are desirable or undesirable aspects of theories. For the purpose of this chapter, there are in particular, two different issues that we will look at to distinguish between different types of theory.

The first issue is our ontology, our philosophy with respect to the nature of reality. To some, reality is governed by a set of rules of how variables inter-relate and science aims to uncover these rules so that we can understand and describe, through our theories, an objective reality that exists independent of us. To others, reality is subjective, a social construct, changing depending on who views it and existing only in our minds as our constructs (Cambridge Dictionary of Philosophy, 1995).

While these positions are obviously extremes and there are intermediate positions, for example to believe in an objective reality as far as the natural sciences are concerned, but that people cannot, for various reasons, be the subject of the same kind of rules, or make exceptions for personal faith, where a God can over-ride any natural laws, this classification is useful for conceptual clarity.

The second, related issue is our epistemology, our philosophy about the nature of knowledge. To some, knowledge is objective, independent of the 'knower' and his or her perspective, attachments and values. To others, knowledge is subjective, formed by the knower(s), reflecting his, her or their viewpoints, attachments and values. (Cambridge Dictionary of Philosophy, 1995).

Obviously, there is a substantial gap between an objective knowledge about an objective reality on one hand and a subjective knowledge about a subjective reality on the other. As we will see, scientific theories that we use for research into construction management are based on an ontology assuming an orderly reality that can be uncovered and known through research and an epistemology of objective knowledge that exists independently of the knower. The kinds of theory, which specify the rules for how variables interact, are often referred to as positivist or post-positivist, or in the recent debate as quantitative research. It utilises information that can be observed and measured and is used to test theories.

The latter, which we will not discuss here, is qualitative research. There is a wide range of opinions about what, exactly, constitutes Qualitative Research. The current use of the term ranges from the use of non-quantitative data to constructivism and critical theory and includes grounded theory (see Chapter 8). The common factor is that the aim of research is not to test or construct theories. Rather, the aim is to understand, to see the world as their subjects see it.

In constructivism and critical theory, qualitative researchers adopt an ontology and epistemology that sees reality as a social construct and knowledge as individual and context dependent. This means that there can be no theories as there are no rules about reality to uncover and there can be no generalisations because knowledge exists only in the mind of the knower (Guba and Lincoln, 1994; Meyer, 1999; Krauss, 2005; Crotto, 1998; Plack, 2005). In fact, there can be no science because theories, forecasting and generalised explanations are the essence of science. Grounded theory, which accepts the existence of theories, but for some reason is normally classified as qualitative, is not a theory about research, but rather a strategy to refer to the results of an empirical research project – a set of inductive research propositions – as a theory, without going through any of the rigmarole of testing and verification that is normally associated with the formation of new theories. For the reader interested in readings promoting qualitative research, the references cited in this paragraph are a good start.

Whatever philosophy we have of science, it is not a given, but drawn from concepts that have evolved over the last three thousand years or so, since we started thinking about science, and it continues to evolve (Runeson and Skitmore, 1999).

Scientific theories

Before we go any further, we need a definition of this kind of scientific 'theory'. In its simplest forms, it consists of a set of assumptions or statements, from which it is possible to logically deduct theorems. These theorems will convert a set of observations into explanations or predictions that can then be compared to real events. In economics, for instance, we have a theory based on a set of assumptions including that sellers want to maximise their profit, that buyers want to maximise their utility, that the addition of an input – like labour – into production increases output but at a diminishing rate if there is no change to the input of other factors and that increasing the consumption of a good increases the wellbeing of the consumer, but at a diminishing rate. From this theory, we can logically deduce that as the price of a good falls, the consumption will increase. This prediction can then be tested against reality and we can get an idea of the usefulness of our theory (Runeson and Skitmore, 1999).

In addition to the theory, we normally have a set of auxiliary statements, statements that sets the environment or the circumstances of our observations. In the example above, the number of sellers and buyer that interact in the market is important for the exact outcome of a change in parameters. With one buyer and one seller, the outcome of a disruption in a market will be quite different to what it would be in a market with many buyers and sellers, so we need auxiliary statements about the nature of the market (Melitz, 1965).

This kind of theory is referred to as 'hypothetico-deductive' theory: we can deduce our theorems from a set of assumptions (statements, hypotheses, axioms) that are supposed to have universal validity. The theorems do not create any new knowledge, because all knowledge was already implied in the original theory. All the theorems do is to present this knowledge in a more specific form.

The alternative is the inductive theory. After watching a great number of white swans, we feel entitled to induce that all swans are white. This is not logically derived, because only when we have seen all swans in the past, present and future can we logically justify this conclusion. On the other hand, by stating that all swans are white, we create new knowledge, although in this case, it happens to be less than 100 per cent correct. In practice, it is difficult to see examples where either of these strict rules of reasoning has been applied in isolation in the formulation of theories. Normally, we have an interaction between deduction and induction.

The final requirement, which we will discuss in detail later, is that the theory should be testable, at least in principle. This means that our theories should be positive theories where a disagreement can be resolved by reference to facts. The alternative is metaphysical theories, which include normative theories, theories about what things ought to be. For obvious reasons, there are some theories that we will rarely or never be able to test, such as our various theories about the origin of the universe. This is why we say 'in principle'. Should a new universe be created, it is possible that we could do the measurements that would test the theories.

Working as a scientist

Now we have a theory, and the next question is: What do we do with it? If you are a Masters or a PhD, whether in the built environment or some other area, you are required to initiate, plan, execute and report on a project where you can demonstrate that you can work as a scientist. This means taking a theory and using it as a scientist would. This pushes the question back one step, to: What do scientists do? Essentially they do three things: develop theories, test theories and use theories to solve problems. According to Popper (1959, 1972, 1982, 1983b), the most important philosopher of science in the twentieth century, the proper job of a scientist is to test theories, and higher degree assessors and academic research fund administrators agree. Using theories or developing technology, while sometimes highly skilled, is not science as such and will not result in a PhD or research funds although it may qualify for a professional doctorate. Scientists may solve professional problems, but not when they are acting as scientists in the sense Popper uses the word. The reasoning that Popper used to derive his idea of the proper job for scientists was based on the dichotomy between verification and falsification and the possibility that any number of theories may produce theorems that are similar.

The dichotomy is easy to demonstrate. Any number of observations of white swans doesn't prove that the statement 'All swans are white' is true. However, one single observation of a black swan is proof that it is false. The possibility that there are many theories that may produce similar theorems is a little bit more difficult to demonstrate, but the concept of an electron is one recent example. Not all that long ago, the atom had a nucleus, orbited by a number of electrons. The electrons were like mini-planets orbiting a mini-sun. In contrast, now, while the electrons are still orbiting around the nucleus, they do not have fixed orbits. Instead, they move from one orbit to another but without passing the space between, dematerialising in one place and materialising in another as the atom releases and absorbs energy. Sometimes, the electrons behave like bullets, at other times they are waves, being both substance and non-substance. When they are waves, they move in a medium that does not conceivably exist, not even a vacuum as the electrons occupy a multidimensional space.

The reason for these changes in the theory is that testing has demonstrated the need for modifications. Both theories can still answer all the old questions, but new developments demand that the theory can answer new questions. To Popper such modifications do not prove that the new theory is right, they don't even reduce the number of potential alternative theories, but many have taken issue with him on that point. They point out that our confidence in a theory increases if attempts to falsify it have been unsuccessful. The theory may not be demonstrated, logically, beyond doubt to be true, but it is the best theory around and the evidence is that it is the one that the scientists use. As Putnam (1974, 1978), for instance, has pointed out: science would be rather meaningless unless it does help us to select the best theory.

While we cannot know if a theory is true, we know that a new version is better than the previous if it can answer the same questions as the earlier theory plus some additional questions. In this way we can see science evolve over time. As we saw with the theory of the atom, the change may be substantial over time, but essentially, it is incremental changes aggregated over time.

The plot gets complex

So far, it has been reasonably simple: one concept of science, one philosophy, a standard set of rules. We shall now introduce a few complications. When we are talking about theories in the natural sciences, we are talking about physical entities, electrons, molecules, electric charges and the like which react with other variables in an exact cause–effect pattern. The cause and effect are direct. The electrons do not stop to think about what to do, or how much, before they do it. However, people do just that (Rosenberg, 1994).

Construction management is mostly about people, interacting with other people. This spans several of the so-called social sciences, like psychology, economics and sociology. In most of the social sciences, the theory is based on an assumption about a motivation. In economics, it is about maximising profit or wellbeing. The producer that wants to maximise profit must think before he or she changes the level of production when the market changes and so must the buyer that wants to maximise his or her wellbeing. They must interpret each situation and make up their minds about what behaviour will be most beneficial.

In these cases, we cannot have falsification in the way Popper stipulated (1959, 1972, 1983a), and for a long time we were not certain whether the social sciences were sciences, and if they were, how we could modify Popper's criterion: *if it can't be falsified, it isn't science* (Hausman, 1985, 1989; Klant, 1984; de Marchi, 1988 or Redman, 1991). Popper was to take almost 50 years before he rejoined the debate he had started on this issue.

In the meantime, Lakatos, one of Popper's former students, published his work, where he accepted theories with an unfalsifiable, metaphysical core, protected and isolated from falsification by a protective belt of derivative theories. Lakatos (1970,

1971, 1977) referred to such a body of theoretical work as a research programme and preferred corroboration rather than falsification (see Backhause, 1994).

Rather than falsification, Lakatos looked at the (scientific) usefulness of a research programme. A theory is a progressive (i.e. useful) theory if it generates scientific progress and regressive if there is no such progress. Since many of the social sciences attempt to explain a wide range of phenomena on the basis of a small number of behavioural assumptions, this evaluation of theories is much more relevant to the actual practice in most of these sciences (Hands, 1993). In fact, in all areas where Lakatos differs from Popper in respect of social science research, he is closer to current practices and has therefore been more influential than Popper, although Popper has retained his status in the natural sciences.

While progressive and degenerative theories can only be recognised in hindsight, the concepts provide a useful way of thinking of theories. Even if degeneration does not necessarily suggest anomalies of the kind required for falsification, and is not necessarily irreversible, a reversal would normally require substantial changes to the theoretical framework (Riggs, 1992).

Finally, after nearly 50 years, Popper modified his criterion for the social sciences:

> ... *as long as a metaphysical theory can be rationally criticised*, I should be inclined to take seriously its implicit claim to be considered, tentatively, as true. (1982, p. 199) and ... Any critical discussion of it will consist, in the main, in considering how well it solves its problems; how much better it does so than various competing theories; whether it does not create greater difficulties than those which it sets out to dispel; whether the solution is simple; how fruitful it is in suggesting new problems and solutions; and whether we cannot, perhaps, refute it by empirical tests. (1982, p. 200). Furthermore, ... *the so called method of science consists in this kind of criticism.* Scientific theories are distinguished from myths merely in being criticisable, and in being open to modifications in the light of criticism (1983b, p. 7).

That resolves the issue of whether the social sciences are really sciences. All we need to do is to accept that we can have causal relationships that are not deterministic (The Cambridge Dictionary of Philosophy, 1995). It does, however, leave us with another serious problem. In the natural sciences, the assumptions are improved as we go on testing them. In the social sciences, because we use motivational variables that cannot be falsified, we cannot improve our motivational assumptions by testing them. As the assumptions cannot be improved, neither can the systems of theorems about social phenomena (Rosenberg, 1994). Hence, there can be no progress in the way the natural sciences progress through empirical testing and attempted falsifications. It does not mean that we cannot have progress, it just means that it is that much harder.

Testing social science theories

If the proper job of the social scientist cannot be attempting falsification, what does the social scientist do? Putnam (1974, 1991) has suggested what he calls three Schemas (Table 7.1).

Schema 1	Schema 2	Schema 3
Theory	Theory	Theory
Auxiliary statements	Unknown	Auxiliary statements
Prediction	Facts to be explained	Unknown

Table 7.1 Putnam's schemas showing legitimate work for scientists.

Schema 1 is the one suggested by Popper, the attempted falsification of a theory. Schema 2 is to establish under what auxiliary assumption a theory can explain an observed situation, as for instance in the discovery of penicillin, when, unexpectedly, a culture of bacteria was found not to grow in the presence of some mould. Schema 3 is when we do not actually know what the outcome is. Whether it is testing a new vaccine for side-effects or establishing how fast the universe is expanding, we have to create the way in which we see or measure reality.

According to Putnam (1974, 1991), all three Schemas are legitimate work for the scientist. All three increase our knowledge and move the sciences forward, in the way Popper envisaged. This tells us a couple of things about scientific progress. It is evolutionary and it involves a large number of scientists, a scientific community. While we know of people like Newton, Einstein or Hawking, few scientists have developed a theory entirely on their own. Rather, if they have 'seen further than others' it is 'because [they] stood on the shoulders of giants' as Newton expressed it. Every individual makes a contribution in increasing our belief that the theory is true whenever they try to falsify a theory without success, define the scope of a theory whenever they test it in a new application or contribute to a modification of a theory whenever they are successfully falsifying a theory. Now and then, when most of the evidence is there, someone provides the missing link and gets his/her name into the history books as they integrate this knowledge into a consistent framework.

However, others, especially Kuhn (1970, 1977), have pointed out that while this is often an appropriate description of progress of science, we sometimes see a totally different development. In the words of Kuhn, we have periods of normal science when the current theory is answering all questions. This is sometimes followed by a scientific crisis when the theory fails to answer a question that is central to the science. In some cases, this is resolved by a modification to the theory and we go back to a period of normal science again, but sometimes this crisis ends in a scientific revolution where everything changes – assumptions, auxiliary variables, definitions and focus. Some of these instances are even known as revolutions outside the scientific community as, for instance, the Copernican or the Darwinian revolutions. In the Copernican revolution, we went from a finite earth-centred, to an infinite sun-centred universe and in the Darwinian revolution from creation without evolution to evolution without creation. Sometimes the revolution is very slow and the decision to change theory can be difficult. Copernicus' theory produced less accurate forecasts and was more complex than the theory it replaced and it took a hundred years until it was widely accepted. Sometimes the decision to accept a new theory has been based on totally non-scientific concerns (Koestler, 1964).

The reason why we have such problems selecting between theories is that testing theories is not a straightforward matter. There are many reasons why we get an inappropriate prediction from a theory and most of them say little about the theory itself, although some of them say a lot about the scientist that does the testing.

The processes involved in testing a theory are outlined in Figure 7.1. The circles in the diagram represent processes that may cause the theory to produce the wrong answer, but as we will see, it is easier to blame the researcher than to reject the theory when things go wrong (Runeson, 1983). Outright rejections of theories are rare. If the logic is not there, obviously the theory fails and should be rejected but that is the only case of automatic rejection we have. Exogenous variables, variables that are not included in the model but may have an impact on the outcome will give the wrong prediction but are, in themselves, not a cause to reject the theory. The same applies to an unrepresentative sample. It may lead to an incorrect prediction even when the model is 'true'. If the measurements used in the test are not appropriate, any testing would be meaningless.

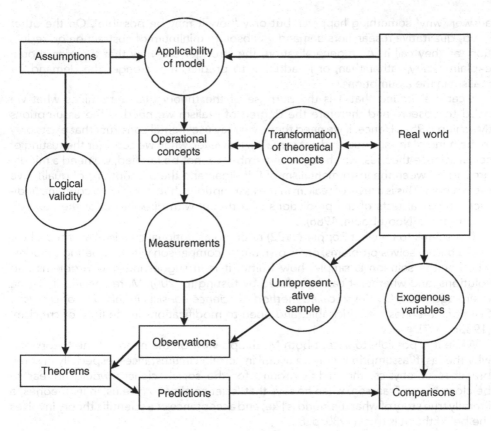

Figure 7.1 A diagrammatic representation of theory testing. The circles represent processes that have the potential to cause distortions of the predictions (adapted from Runeson, 1983).

The applicability of the model is important like the correspondence. Both are difficult and complex concepts, theoretically, philosophically and practically. If the model is not designed for the context in which it is applied, it will not work, but there is no automatic cause to reject the model, and we will discuss this later. The correspondence concerns how well the theoretical concepts of the theory are translated into operational concepts, which can be observed and measured. While theoretical concepts are normally clear, there is seldom an easy conversion into observable concepts. This applies not only to things like productivity or safety, but also to seemingly simple items like costs or construction delays.

We will look first at some of the issues that are important for the applicability of models, before coming back to correspondence rules. The first issue we will look at is the realism of assumptions. The more realism we have in the assumptions, the easier it is to determine if a theory is applicable. Most introductory texts in the social sciences start by explaining that reality is too complex to observe and that some abstractions are necessary in all theorising, so that we can focus on the essential actors and relationships. However, there is no agreement about how much 'abstracting' is desirable. At one extreme, we have people like Friedman (1953), who see no need for any form of realism in the assumptions. He even suggests that assumptions can be seen as 'as if' statements. Things happen as if the assumptions were true. This means that the assumptions may not be supposed to be true and that a theory does not aim to answer 'why' something happens, but only 'how it may be possible'. On the other hand, qualitative researchers demand an absolute minimum of abstraction or 'reduction' as they call it. As a generalisation, the stronger the view that theories should explain reality, rather than, or in addition to predict, the stronger the demand for realism in the assumptions.

It can be argued that it is the purpose of the theory that determines what we need to observe and therefore the degree of realism we need in the assumptions (Machlup, 1952). Hence, if we need them, we may use mental construct that exists only in the mind of the scientist and not in reality. However, when we consider the testing of social science theories, which as you remember cannot be falsified, we need a middle ground between the instrumentalism of Friedman and the descriptions of qualitative researchers. This is a level of realism in the assumptions that allows us to combine (indirect) empirical tests of the predictions of a theory with direct tests of the auxiliary assumptions (Nooteboom, 1986).

This seems to be what Popper (1982) refers to as 'rational criticism' as it tests how well a theory solves problems, on its own and in comparison with competing theories, whether the solution is simple; how fruitful it is in suggesting new problems and solutions; and whether it can be refuted by testing (p. 200). More specifically, as he later suggested, '... the so called method of science consists in this kind of criticism. Scientific theories are criticisable, and open to modifications in the light of criticism' (1983b, p. 7).

While it is possible to accept both Machlup's view on the needs of the theory and why the 'as if' assumptions may be useful in certain circumstances, especially in some branches of physics, the middle-ground for the social sciences would appear to be close to van Fraassen, when he says that 'science aims to give us, in its theories, a literally true story of what the world is like, and acceptance of a scientific theory involves the belief that it is true' (1980, p. 8).

We need realistic assumptions, because a theory, by itself, has no implications that we can observe, and we cannot accept it in isolation from other statements. The theory 'if A is true, then B is true' does not tell us anything about the world unless A is true. Only when we combine the statements 'if A is true then B is true' and 'A is true' does it tell us that B and all implications of B are true (Melitz, 1965; Putnam, 1974). This means that there is a strong logical argument that all assumptions should be true. The closer the assumptions are to reality, the better we can identify uniquely the domain of the theory where it applies. Without this, a theory cannot be tested, as it can be argued that any falsification is because the theory has been incorrectly applied.

It is worth noting, however, Friedman's (1953, pp. 19–36) argument that, whilst assumptions may be used to specify the domain of a theory, it does not necessarily mean that they can be used to define uniquely, the domain, as he claims that '... there is no inconsistency in regarding the same firm as if it were a perfect competitor for one problem and a monopolist for another'. However, this extreme view is not widely shared.

How variables are measured is crucial for all testing. The hypothetico-deductive models that we are discussing here can only be related to reality through correspondence rules that translate the concepts in the theory into corresponding concepts in reality. Similarly, inductive theories require that the observational variables can be transformed into unique theoretical concepts. This correspondence, through which a theory relates to the external world, is essential for all uses and all testing of a theory. However, the correspondence rules are seldom included in the theory, and for all cases where they are not, there is a problem.

The problem is best summarised in the so-called 'Duhem-Quine thesis', which states that when absolute correspondence rules are not specified in the theory, the empirical testing of the theory can never be conclusive (Bechtel, 1988). This is because operational variables cannot be taken as absolute and un-modifiable parts of the theory. They must be seen as tentative and subject to revision as we learn more. When a theoretical prediction is threatened by an empirical result based on operational definitions of theoretical terms, one way to protect the theory is to immunise it by requiring revised operational definitions.

Some interpretations of the Duhem-Quine thesis are even more radical, and suggests that in some circumstances:

> in deciding where to modify our theoretical structure in the face of negative evidence, we may choose to modify the propositions of logic and mathematics as well as those more generally thought of as part of empirical science. (Bechtel, 1988, p. 43)

Whether we accept the Duhem-Quine theorem or not, the influence of this kind of thinking is reflected in the many cases in the social sciences where theories that are obsolete and should have been rejected many years ago are still in circulation, ready to trap any researcher who is not prepared to spend the necessary time to set up the theoretical framework and conceptual model.

A solution (or two)

We now have two good reasons why we cannot falsify a theory – the motivational assumption and the Duhem-Quine theorem – and it is also logically impossible to prove a theory. Clearly, we cannot be scientists in the way Popper initially suggested, but we can still do a few things to help us develop and promote good theories. We can corroborate them by repeatedly testing the theories and finding that they seem to be true. As Putnam suggested, we can then induce from this process that the theory is likely to be adequate. We may develop new theories on the foundation of the research programme that Lakatos suggested was the evidence of a progressive research programme.

We can compare how closely the model resembles the situation we are investigating. This concept of testing has been endorsed by van Fraassen. While we cannot know for certain, we may reasonably believe that a model is empirically adequate when all the aspects of the theory correspond to the situation where it is applied (van Fraassen, 1991, p. 193). By extension, a model is empirically inadequate without this fit.

We may also use adduction. We have the theory 'if A is true, then B is true; A is true, therefore B is true' where we must demonstrate that A is true (*affirming the antecedent*). If we change the statement to read 'if A is true, then B is true; B is true'; it does not follow that A is true. To assume that A is true would be to *affirm the consequent*, and the conclusion that A is true is not a logical necessity. This kind of reasoning belongs more to the realm of politics than science. Adduction changes the emphasis: 'B is unexpectedly observed. If A was true, B would be true. Hence it is likely that A is true' (Hoover, 1994, p. 301, following Peirce). This is clearly not a logical deduction, as it is logically invalid, but rather a form of inference. It is also a form of empirical testing of a theory as it indicates the extent to which observations fit a theory.

Building new theories

With so many potential problems with testing theories, it would be reasonable to ask if we cannot do scientific research without this. The answers are 'yes' and 'no'. There are still areas where there are no theories although, more often than not, a failure to find an appropriate theory for a special phenomenon has more to do with sloppy research than a lack of theories, so it is nowhere near as common as research proposals from new PhD candidates would make us believe. Here, of course it is the legitimate work of scientists to formulate new theories rather than test existing one's.

However, the procedure when we aim to develop a new theory is very much the same as when we test an existing theory. In a research project, we have a cycle that goes from theory to hypothesis to data collection to analysis and back to theory for modification or corroboration. If the aim of the research is theory formation, we face the same cycle but start at data collection and end with a hypothesis offered for testing of the theory we have proposed. In other words we do not escape from the theory as such, unless our philosophy makes us believe that there can be no theories, as in some qualitative research.

Conclusions

As with other scientists, built environment researchers need theories for their work to be understood, communicable, and ultimately implemented for the benefit of society. This paper has brought together and discussed a number of issues that are important in both the formulation and testing of scientific theories. The aim is to provide a broad overview over an essential, but quite complex topic, and the emphasis has been to demonstrate the connection between theories and our philosophy of the world and our knowledge of it. This relationship is particularly important when we are involved in the kind of research that leads to higher degrees or academic research grants. We have also suggested that the major difference between this type of research and the so-called qualitative research is not in the measurement of the information but in our fundamental perception of the world.

While we have demonstrated that the 'traditional' idea of falsification cannot be applied in the social sciences that contribute to construction management, we have pointed to alternatives in the form of corroboration, including adduction.

References

Backhause, R.E. (1994) *New Directions in Economic Methodology*, Routledge & Kegan Paul, London.

Bechtel, W. (1988) *Philosophy of Science*, Lawrence Erlbaum Associates, Hillside, NJ.

Cambridge Dictionary of Philosophy (1995) *Social Science*, Audi, R. (ed.), Cambridge University Press, Cambridge.

Crotto, M. (1998) *The Foundations of Social Research: Meanings and Perspectives in the Research Process*, Sage, Thousand Oaks.

de Marchi, N. (1988) Popper and the LSE Economists, in de Marchi, N. (ed.) *The Popperian Legacy in Economics*, Cambridge University Press, Cambridge.

Friedman, M. (1953) *Essays in Positive Economics*, Chicago University Press, Chicago.

Guba, E. and Lincoln, Y. (1994) Competing paradigms in qualitative research, in Denzin, K. and Lincoln, Y. (eds) *Handbook of Qualitative Research*, Sage, Thousand Oaks, pp. 105–118.

Hands, D.W. (1993) *Testing, Rationality and Progress: Essays on the Popperian Tradition in Economic Methodology*, Rowman & Littlefield, Lanham, MD.

Hausman, D.M. (1985) Is Falsification Unpractised or Unpracticable, *Philosophy of Science*, 15, 313–319.

Hausman, M.D. (1989) Economic Methodology in a Nutshell, *Journal of Economic Perspectives*, 3(2), 115–127.

Hoover, K. (1994) Pragmatism, pragmaticism and economic method, in Backhause, R.E. (ed.) *New Directions in Economic Methodology*, Routledge & Kegan Paul, London.

Klant, J. (1984) *The Rules of the Game: The Logical Structure of Economic Theories*, Cambridge University Press, Cambridge.

Krauss, S. (2005) Research paradigms and meaning making: A primer, *The Qualitative Report*, 10(4), 758–770, also available at http://www.nova.edu.ssss/QR/QR10-4/krauss.pdf.

Koestler, A. (1964) *The Sleepwalkers, A History of Man's Changing Vision of the Universe*, Penguin Books, Harmondsworth, Middlesex.

Kuhn, T.S. (1970) *The Structure of Scientific Revolutions* (2nd Edition), University of Chicago Press, Chicago.

Kuhn, T. (1977) Second thoughts on paradigms, in Suppe, F. (ed.) *The Structure of Scientific Theories* (2nd Edition), University of Chicago Press, Chicago, pp. 459–482.

Lakatos, I. (1970) The methodology of scientific research programs, in Lakatos, I. and Musgrave, R. (eds) *Criticism and the Growth of Knowledge*, Cambridge University Press, Cambridge.

Lakatos, I. (1971) History of science and its rational reconstruction, in Buck, R.C. and Cohen, R.S. (eds) *Boston Studies in the Philosophy of Science*, 8, 91–136.

Lakatos, I. (1977) *Proofs and Refutations*, Cambridge University Press, Cambridge.

Machlup, F. (1952) *The Economics of Sellers Competition*, John Hopkins Press, Baltimore.

Melitz, J. (1965) Friedman and Machlup on the significance of testing economic assumptions, *Journal of Political Economy*, 73, 37–60.

Meyer, P. (1999) *An Essay on the Philosophy of Social Science* (revised 31 January 2001), available at http://www.serendipity.li/jsmill/pss2.htm.

Nooteboom, B. (1986) Plausibility in economics, *Economics and Philosophy*, 2(2), 197–224.

Plack, M.M. (2005) Human nature and research paradigms: Theory meets physical therapy practice, *The Qualitative Report*, 10(2), 223–245.

Popper, K. (1959) *The Logic of Scientific Discovery*, Hutchison, London.

Popper, K. (1972) *Conjectures and Refutations: The Growth of Scientific Knowledge*, Routledge & Kegan Paul, London.

Popper, K. (1982) in Bartley III, W.W. (ed.) *The Postscript to the Logic of Scientific Discovery*, Vol. 3, Rowman & Littlefield, Towowa, NJ.

Popper, K. (1983a) *Realism and the Aim of Science*, Hutchinson, London.

Popper, K. (1983b) in Bartley III, W.W. (ed.) *The Postscript to the Logic of Scientific Discovery*, Vol. 1, Rowman & Littlefield, Towowa, NJ.

Putnam, H. (1974) The 'corroboration' of theories, in Schilpp, P. (ed.) *The Library of Living Philosophers*, Vol. XIV, *The Philosophy of Karl Popper*, Open Court Publishing Company, LaSalle, IL, pp. 221–240, reprinted in (1991) in Boyd, R., Gasper, P. and Trout, J. D. (eds) *The Philosophy of Science*, The MIT Press, Cambridge, Mass., pp. 121–136.

Putnam, H. (1978) Retrospective note (1978) a critic replies to his philosopher, reprinted in Boyd, R., Gasper, P. and Trout, J.D. (eds) *The Philosophy of Science*, The MIT Press, Cambridge, Mass., pp. 136–137.

Redman, D.A. (1991) *Economics and the Philosophy of Science*, Oxford University Press, Oxford.

Riggs, P.J. (1992) *Whys and Ways of Science: Introducing Philosophical and Sociological Theories of Science*, Melbourne University Press, Melbourne.

Rosenberg, A. (1994) What is the cognitive status of economic theory, in Blackhouse, R.E. (ed.) *New Directions in Economic Methodology*, Routledge & Kegan Paul, London.

Runeson, G. (1983) *Economics of Building*, Victoria University Press, Wellington, NZ.

Runeson, G. and Skitmore, M. (1999) *Writing Research Reports*, Deakin University Press, Deakin.

van Fraassen, B. (1980) *The Scientific Image*, Oxford University Press, Oxford.

van Fraassen, B. (1991) *The Pragmatics of Explanation*, in Boyd, R., Gasper P. and Trout, J.D. (eds) *The Philosophy of Science*, MIT Press, Cambridge, Mass.

Chapter Eight

Grounded theory

Kirsty Hunter and John Kelly

Introduction

Grounded theory is a method of research in which theory is derived from a structured data set with or without a preliminary research question. This chapter introduces the work of Glaser and Strauss, who first introduced the methodology in 1967, and describes its evolution in the context of other theory building methods. The evolution of the grounded theory methodology is reflected in the post-1967 differences in approach between Glaser, who adopted a pure form of the derivation of theory from data and Strauss and Corbin who advocated an approach driven by a research question. Methods for data collection and data processing are introduced and an explanation given of the analysis of data. The challenges of using the grounded theory methodology, its scope and limitations are discussed before concluding by correlating the principles of grounded theory with the importance of triangulation and the characteristics of good theory development.

What is grounded theory?

Grounded theory is a methodology which involves a systematic process of gathering and analysing a finite set of data to evolve a theory based upon the data. The theory may then be used to predict and explain phenomena. The introduction of grounded theory as a research methodology is generally attributed to Glaser and Strauss (1967) whose work has spawned a number of different versions. The term 'grounded' is used because the theory is drawn from the data and not from speculation or preconceived ideas, allowing the practitioner to have more control and understanding of situations (Glaser and Strauss, 1967). Allan (2003) states that: 'if the data has been analysed without a preconceived theory or hypothesis, that theory is truly grounded in the data because it came from nowhere else'. A theory is built as opposed to being tested and offers an explanation of phenomena rather than just a set of findings (Strauss and Corbin, 1998).

Glaser and Strauss (1967) state that grounded theory is 'derived from data and then illustrated by characteristic examples of data'. This is supported by Strauss and Corbin (1998), who define theorising as 'the act of constructing from data an explanatory scheme that systematically integrates various concepts through statements of relationship'. Coffey and Atkinson (1996) suggest that theorising involves the identification of patterns and relationships. One of the fundamental benefits of using grounded theory is that it increases the knowledge base by developing new theories (Heath and Cowley, 2004).

Grounded theory is at the interpretivist, post-positivist end of the philosophical continuum (Scott *et al.*, 2002; Backman and Kyngas, 1999). The continuum in Figure 8.1 shows the differences between approaches. It involves the researcher having no

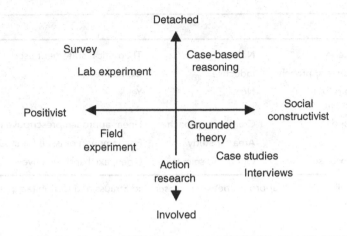

Figure 8.1   Research methods are driven by approach (Kelly, 2004).

preconceived ideas of results and does not begin with a theory to be proved but an area to be studied, allowing for the theory to emerge. Sources for generating theory vary; these include case studies, interviews, historical accounts, field observations and documents. Grounded theory differs from other methodologies in the sense that it involves theory development (Denzin and Lincoln, 2000). It has also been described as 'a reaction against more traditional scientific approaches' which include hypothesis testing and quantitative methods (Cronholm, 2002). Allan (2003) highlights that grounded theory is different from other methodologies in the sense that a concurrent approach is taken by both collecting and analysing the data.

Having proposed the research methodology, Glaser and Strauss subsequently disagreed over its precise application which led to three different approaches to grounded theory (Esteves et al., 2002). These are: Glaser and Strauss (1967), Strauss and Corbin (1990) and Glaser's (1978, 1992) interpretation. The principal distinction between approaches is that Glaser predicates that there should not be a conceived theory in mind, whereas Strauss and Corbin (1998) highlight the requirement for a theoretical statement to enable an explanation or prediction of theory. Table 8.1 highlights the different characteristics in approach between Glaser, and Strauss and Corbin.

The researcher should be clear on the approach adopted for the grounded theory study. Table 8.2 outlines the characteristics of the various research approaches that may be adopted.

Glaser has criticised Strauss for making assumptions from the data instead of looking at what exists in the data (Heath and Cowley, 2004). If a combination of approaches is used, it is recommended by Heath and Cowley (2004) that 'boundaries between the two should be maintained rather than a synthesis attempted'. Therefore, the researcher should be aware of the characteristics of the research approach used for the research study.

Substantive to formal theory

Whetten (1989) states that it is very unlikely for a new theory to be built from first base and proposes that new theory is built on existing theory in order to make

Characteristic	Glaser	Strauss and Corbin
Preconceived theory	No	Theoretical statement used
Inductive/deductive approach	Inductive	Deductive
Use of literature prior to theory development	No	Yes
Technique focused	Creative approach	Linear approach prescriptive in technique
Level of detail	Area for study	Phenomenon or issue for study
Coding: open, axial, selective	Open and selective	Open, axial and selective

Table 8.1   The differences in approach between Glaser, and Strauss and Corbin (adapted from Hunter et al., 2005).

Characteristics	Examples of approach
Research strategy/technique	Case studies, field observations, archival records, interviews.
Use of software package for data sorting and/or analysis	A manual analysis without use of a specific content analysis software package or alternatively a software package such as NUD * IST or askSAM may be used.
Outcome of approach	Theory generation/theoretical framework.
Inductive/deductive	Inductive using the Glaser approach, and deductive using the Strauss and Corbin approach.
Assumptions made	Yes, a research proposition/theoretical statement is used (Strauss and Corbin approach). No preconceived theory in mind (Glaser approach).
Prior knowledge	Literature and experience may be used to inform study if the Strauss and Corbin approach is adopted.
Approach chosen	Glaser, or Strauss and Corbin, or combination of both.

Table 8.2   Characteristics of the research approach.

improvements. Theory may be described as substantive or formal. Substantive theory uses raw data and focuses on one particular area which restricts theory application to that specific area (Glaser and Strauss, 1967). Formal theory is developed from a substantive base but generated from findings from many studies and therefore is broader in scope. Coffey and Atkinson (1996) describe formal theory as being more generic.

Theory is generated from a combination of literature, observations, common sense and experience. The use of literature is important in theory building as it allows the researcher to build their case through supporting evidence (Eisenhardt, 1989). However, various authors in the field of grounded theory debate the impact of literature review if conducted prior to the study. Some authors suggest that a review of the literature should be done after the theory is developed to ensure that the theory is grounded in the data and to confirm the theory, whereas others suggest a review of literature prior to the study which will identify any gaps in the knowledge (Cutcliffe, 2000). Disagreement exists between Glaser and Strauss who each have different views on the role of the literature during the course of the study. This is identified in Table 8.1.

Eisenhardt (1989) stresses that no theory is derived from the ideal situation, of a 'clean theoretical slate' defined as an absence of assumptions or presumed theory at the commencement of the study that could result in a theory which is limited and biased. This is supported by Selden (2005) and Heath and Cowley (2004), who highlight that no one can be completely cut off from their own experiences and reading which will undoubtedly influence the process in some form. They suggest that focused reading can be done to allow the literature to support the developed theory. It has also been suggested by Goulding (2005) that literature is consulted as part of an iterative, inductive and interactional process of data collection. Wacker (1998) makes reference to the literature to ascertain what concepts are worthy of further investigation. This indicates that the literature may be used in a variety of ways depending on the approach chosen by the researcher.

Data collection and analysis

Three main categories of data are used in grounded theory research (Douglas, 2003). These are: field data (notes), interview data (notes, recordings, transcripts) and any other existing literature. For example, the researcher may use case studies or conduct interviews in their study which involve qualitative data. Data available in case studies or interview transcripts should be sufficiently rich to generate theory. The number of cases or interviews used in generating theory is not an important consideration when generating theory (Esteves et al., 2002; Glaser and Strauss, 1967). This is because a range of two to a great number of cases or interviews can confirm a theory dependent on the data available. Therefore, it cannot be stated at the outset, how many case studies or interviews will be required, this can only be determined at the end of the study. Glaser and Strauss (1967) emphasise that, as theory grows, it becomes reduced, meaning that there is little to add to the core theory once a theory has been developed from the data, and therefore after a certain point of theory building it will be unnecessary to obtain more case studies or conduct more interviews.

Strauss and Corbin (1998) describe the process of data analysis as: 'the interplay between the researcher and the data'. Eisenhardt (1989) states that central to theory building is analysing data. However, it is stated that; 'a huge chasm often separates data from conclusions'. Miles and Huberman (1984) state that: 'One cannot ordinarily follow how a researcher got from 3 600 pages of field notes to the final conclusions …' Eisenhardt (1989) makes the point that people do not process information effectively and are quick to draw conclusions. To prevent this, the data should be viewed in different ways. Creswell (1994) highlights that there is no one right way to analyse the data and, therefore, the method that is most suited to the data at hand should be chosen.

Data analysis should be conducted in conjunction with data collection to identify areas of interest early on and to ensure that the method used was well suited. The unit of analysis should also be identified at this stage. Yin (1994) defines the unit of analysis as being related to the research questions, or alternatively, the research propositions.

Analysis may be done using comparative analysis described as: 'a strategic method for generating theory' (Glaser and Strauss, 1967). Coffey and Atkinson (1996) cite Tesch (1990), who has identified at least 26 analytic strategies for qualitative analysis. Comparative analysis usually commences with sorting the properties into their categories and then conducting a line-by-line analysis and noting the common themes (Goulding, 2005). Glaser and Strauss (1967) state that the constant comparative method leads to the reduction of theory over time as the theory is developed. The process of comparative analysis is expected to reveal similarities and differences in the data. The focus for the study should be on 'regularities' in the data and generation of theory will be restricted to the search for regularities in the data (Glaser and Strauss, 1967). The process of data analysis will continue until there is theoretical saturation which occurs when it becomes obvious that similar issues are reoccurring. Saturation is when no additional information can be found and the resulting theory can be challenged (Glaser and Strauss, 1967).

Coffey and Atkinson (1996) suggest that the use of computer software is more suited for storage and retrieval of information than for analysis. The study may involve computer support for only those particular tasks. Methods of sorting the data include colour coding, code words and index cards. Crabtree and Miller (1999) cited in Scott *et al.* (2002), recommend the use of colour codes when there are relatively few codes. The process of sorting makes for an easier transition from the data to the theory as the links are evident in the colours. This is supported by Coffey and Atkinson (1996), who cite Seidel and Kelle (1995): 'codes represent the decisive link between original "raw data" … and theoretical concepts'. They also cite Miles and Huberman (1994), who describe codes as: 'tags or labels for assigning units of meaning'. Therefore, coding makes the analysis easier by identifying the meaningful data for interpretation. Coding is also described as an essential procedure prior to analysing the data (Strauss, 1987 cited by Coffey and Atkinson, 1996).

It is important to investigate the attributes of both manual and electronic methods before a decision is made on the most suitable option. The attributes of both approaches are described in Hunter *et al.* (2005), who describe two different applications of grounded theory: theory generation and a theoretical framework.

The theory building procedure

A procedure for theory building involves defining the variables, specifying the domain, building internally consistent relationships, and making predictions. The same process should be undertaken regardless of the methodology used (Wacker, 1998). Figure 8.2 highlights the process that should be undertaken.

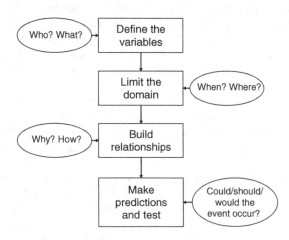

Figure 8.2   A general procedure for theory-building (adapted from Wacker, 1998).

A criticism of some research work that adopts grounded theory is that it fails to report the stages involved. The work of Glaser and Strauss is cited by Eisenhardt (1989), who concludes that they did a lot of work on building theory but not much on how to go about it. Douglas (2003) contends that 'there remains a paucity of published accounts of the application of grounded theory'. Woodhead and Downs (2001) cite Spiggle (1994), who offers a five-step process for implementing grounded theory. The first step involves categorising and coding the data, followed by abstraction to identify the high-order categories. The data are then compared through a process of exploration of the differences and similarities, followed by iteration to review the analysis and, finally, refutation to test the conclusion by finding anything that might challenge or disprove the theory.

Collection, coding and analysis should be conducted as a parallel process. Denzin and Lincoln (2000) highlight that the concurrent process prevents the researcher 'drowning in the data'. This process should continue until there is theoretical saturation. The developmental process of theory building is described by Douglas (2003), who states that 'concepts have been identified, developed, discounted, and merged in order to produce the component concepts of the emergent theory'. Allan (2003) describes the grounded theory process as looking and re-looking for emerging codes.

Data sorting

Miles and Huberman (1984) suggest that tabular displays and graphs are good methods of displaying the data which also prevents the loss of information through a rigorous coding process. As data are accumulated this may present increasing difficulties in analysing the data due to the quantity of data to interpret. This may result in a modification of the process which may involve the construction of 'memos' similar to a key to list all the highlighted properties, the colour code, and the common area. This allows the referencing of the memos to identify the common properties in the subsequent data input when added to the data file, spreadsheet or computer software package. Selden (2005) describes how memos can be used and suggests that diagrams and models can be drawn to indicate the category relations using boxes and arrows. Figure 8.3 is an example extract from a grounded theory study exploring common construction-related issues across various case study projects. An example of three case studies has been given which shows the use of memos to identify the common properties under two categories. Category one is budget/cost/finance and category two is constructability/construction/buildability. An example showing a selection of common properties is evident from the arrows branching from the memos/key across the cells for the three case study projects.

Coding and comparison groups

Glaser and Strauss (1967) state that the rule is to maximise the comparison groups. This may be done by ensuring that as much data as possible has been used to generate the theory. The properties are compared using a process of comparative analysis to search for similarities and differences. A second sweep of coding may be done in reverse order to ensure that the first set of data input was not the only set to be used to compare across the groups. Categories and properties are two different elements of theory (Glaser and Strauss, 1967). The level of detail chosen for identifying categories and properties from the data should be fairly general to ensure that there is openness towards interpreting the data (Scott et al., 2002).

CASE STUDY NUMBER	1	2	3
CASE STUDY REF.	MRR	RSE2	CLV & CLYTWS
MEMOS/KEY	**Budget**	**Cost**	**Finance**
Design cost →	Cost in design terms →	Cost limits →	Cost reflecting design
Cost limits/certainty →		Availability of funding →	Require cost certainty
Funding - availability →		Inflation due to programme →	Inflation
Inflation →		Costs in use →	
Operating costs →			Quality promise
Quality →			Right price for job
Right price →			Understanding project
Project understanding →			Value for money
Value for money →			
Final account →	Final account procedure		Life cycle of materials
Contingencies →	Contingencies		No nasty surprises - open and honest
Other →	Recovery of costs	Aggregate tax	Openness
Other	Preliminaries	Cost in use to X	Process to manage out turn cost
Other	Provisional sums – what's included?	Cost in use to tenant	Sufficiency of tender
Other	Cost		
	Cost plan		

MEMOS/KEY	**Constructability**	**Construction**	**Buildability**
Co-ordination →	Co-ordination of the works →	Services co-ordination →	Defects on existing construction
Defects →	Final Defects →	Phasing of the works →	Potential for use of standardisation
Planning/phasing →	Planning →	Prefabrication →	
Prefab/standardisation →		Quality of labour →	Interfaces of materials / material selection
Quality →	Quality →	Impact of traffic management →	
Traffic management →		Accuracy of services →	Construction time
Services →			Contractor division responsibility
Materials →		Statutory authority agreements	Necessity of some internal work?
Buildability →	Buildability	Avoid compaction of gardens by plant	Maintenance implications
Other	Repeat mistakes	Existing foundations	Dimensional variability
Other	Clerk of works	Existing site service	Innovation vs tried and tested - risk
Other	Precast concrete	On-site waste and off-site tipping	M & E
Other	Deck	Minimise waste	Long term maintenance cost
Other	Problem solving		
Other	Lack of skill		
Other	Procurement		
Other	Protection		

Figure 8.3 Example of the use of memos for identifying common properties across categories.

The coding process in grounded theory is a form of content analysis. There are three different types of coding in grounded theory: open, axial and selective. Open coding involves defining the categories, axial coding entails relating properties to categories, and selective coding identifies the relationships and commonalities that lead to the development of the theory. Douglas (2003) states that creativity occurs when the researcher makes associations that bring fresh insights into the data collected.

Theory development in case studies

The use of case study theory building is considered to be appropriate in new areas or to provide different perspectives. The advantages with the use of case studies in theory building is the close link with reality where the theory is likely to be novel, testable, and be empirically valid (Eisenhardt, 1989). Most research is conducted the other way round from theory to data; theory building begins with data and results in theory (Eisenhardt, 1989). This is supported by Wacker (1998), who states that: 'the establishment of causal relationships are under-researched'. Fact-finding research explores differences in the data whereas theory building research looks at the similarities (Wacker, 1998).

It is recommended that a sample population of case studies that will build on the emergent theory be chosen prior to commencement of the study as a method of control (Eisenhardt, 1989). This contradicts the grounded theory literature which ascertains that the number of case studies should not be limited and only defined once theoretical saturation occurs: that is when nothing new can add to the theory. The principles of grounded theory ensure that the theory is 'grounded' in the data. A theory that is 'grounded' exists when the theory has been drawn from the data and not from speculation or preconceived ideas.

Eisenhardt (1989) outlines that there is a great deal of uncertainty associated with building a theory from case studies. The lack of clarity results from the data being predominantly qualitative which makes it difficult to code. Mintzberg (1979) states that: 'we uncover all kinds of relationships in our hard data, but it is only through the use of this soft data that we are able to explain them'. The soft data referred to is the qualitative data.

Grounded theory challenges

There are challenges associated with grounded theory which should be taken into account when undertaking research. Some common problems are:

- Some theories can be very 'low content', which means that the researcher has developed the theory too early on in the course of the research work (Kinach, 1995; Haig, 1995).
- There tends to be a degree of uncertainty in determining when the analysis stage is complete; the point when theoretical saturation has occurred (Goulding, 2005; Star, 1997).
- There is often a lack of understanding of how the grounded theory is to be used (Burca and McLoughlin, 1996).
- The theory is phenomena and not data interpreted by the researcher and therefore the method is not scientific (Burca and McLoughlin, 1996).
- Researchers make assumptions based on their own personal beliefs and values that will inevitably affect their data (Selden, 2005; Burrell and Morgan, 1979).
- 'Prior theoretical socialisation in a researcher may produce ideational and ideological baggage, which inhibit forming fresh ideas and promotes tunnel vision' (Charmaz, 1990).

- The researchers' disengagement from the data to develop the theory. If this is not done it is suggested that the theory will be 'naive and ill-constructed' (Backman and Kyngas, 1999).

Burca and McLoughlin (1996) state that grounded theorists must accept responsibility for their own interpretations of the data which may be influenced by previous experience and knowledge in the area. Cutcliffe (2000) suggests that researchers should recognise how their knowledge in the area under investigation may have impacted the study and should explain the biases and possible impacts this may have. It is suggested that the researcher questions the theory development by asking questions such as: 'does that thought originate from my knowledge, experience or beliefs or does it belong to the interviewees?'.

Dainty et al. (2000) summarises the three common problems associated with grounded theory analysis as: 'data overload, complex procedures and a lengthy analytical phase'. In terms of data overload, one of the weaknesses of theory building is the amount of data that is studied which can result in a diluted theory which has attempted to contain everything. Another weakness associated with general theory building outlined by Wacker (1998) is that of 'concept stretching', which occurs when the researcher broadens the scope for application of the theory by reading too much into the variables concerned.

It is evident that there is an increasing amount of jargon associated with grounded theory from different authors that can be confusing for researchers (Burca and McLoughlin, 1996). This is supported by Backman and Kyngas (1999), who suggest that the researcher should follow one particular author, that is Glaser or Strauss, and then develop their own method using one of these as a foundation. It is considered that a degree of confusion would occur if the researcher were to apply a combination of applications of grounded theory from different texts and the resulting findings would be lacking in substance. However, a combination of approaches is possible provided the researcher explains how the research study 'fits' with each approach. The approach chosen will ultimately depend on the nature of the study.

Coding data and defining categories and the importance of not forcing data into codes and ensuring that adequate categories cover the data subjects is a challenge cited by Burca and McLoughlin (1996), who describe the area as: 'soft and lacking rigour'. Researchers are warned of the dangers of being interested only in patterns in the data in preference to substance. Common problems in sorting unstructured textual data occurs in grounded theory where the data contains vernacular words (Star, 1997).

Burca and McLoughlin (1996) emphasize that theory is built on the researchers interpretations: 'all interpretations are temporarily limited and therefore fallible'. Heath and Cowley (2004) suggest that the researcher could shape the research to explore the area that they are interested in and therefore draw conclusions that are not wide ranging. Robrecht (1995) cited in Heath and Cowley (2004) outlines the importance of looking at the data rather than for specific data.

Scope and limitations of theory

In considering researcher bias when analysing the data Strauss and Corbin (1998) state that: 'it is not possible to be completely free from bias' and therefore slightly different interpretations of the theory may emerge dependent on the researcher. In addition, prior experiential knowledge on the part of the researcher is hard to contain and may influence the analysis in some way (Edwards et al., 2002) and (Scott et al., 2002). Other considerations to take into account that play a significant role in the resulting theory are; quality of data, access to data, researcher's ability to analyse/interpret data, and researcher's deductions (Selden, 2005; Cronholm, 2002). Any theory generated is constrained by the data available.

In terms of the limitations of external validity it is important to highlight how the process can be repeated to ensure that a similar procedure is undertaken to achieve a reliable theory.

A good theory

A good theory is described as being parsimonious, testable, and logically coherent (Eisenhardt, 1989 cites Pfeffer, 1982). Wacker (1998) adds to Eisenhardt's list with the following test of good theory:

- Uniqueness: means that one theory must be differentiated from another.
- Conservatism: a current theory can only be replaced by a new theory if it is superior in important aspects. This means challenging other theory that opposes it.
- Generalisability: the more areas that a theory can be applied to makes the theory more important.
- Productive: a good theory is one that is fertile in generating new models and hypotheses.
- Parsimony: the fewer assumptions the better.
- Internal consistency: the theory gives adequate explanation for all relationships.
- Empirical risk: the theory should hold itself up for refutation and not hide behind limiting caveats.
- Abstraction: the theory should be independent of time and location.

Poole and Van de Ven (1989) offer a succinct definition of a good theory: 'a limited and fairly precise picture'. This is in the sense that the theory does not encompass everything and instead outlines the theory's scope and limitations. This corresponds with Whetten (1989), who suggests that it is best to include more variables to begin with and then refine later.

Important considerations in grounded theory building are:

- Strength of the method, the evidence grounding the theory, the analytical procedure, supporting evidence and other explanations.
- Enough evidence should support the developed theory to allow others to reach similar conclusions (Eisenhardt, 1989) and the logic in the building of theory should be capable of replication (Eisenhardt, 1989).
- The theory must result in something new where the ultimate goal is new theory.
- The goal of good theory is to establish why relationships exist in the theory and what this could lead to. It is stated that: 'researchers can define theory as a statement of relationships between units observed or approximated in the empirical world' (Wacker, 1998).
- Any inconsistencies should also be identified and not ignored. Inconsistencies may be used for the development of other theories (Poole and Van de Ven, 1989).

Wacker (1998) suggests that theory has four basic criteria: these are conceptual definitions, domain limitations, relationship-building, and predictions (Figure 8.2). The conceptual definitions are the concepts involved in the theory. The domain establishes the when and where of theory application. The relationship aspect of theory building explains how and why concepts are linked. The predictions involve the ability to determine when and where the theory will occur. Wacker (1998) states that the natural language questions of who, what, when, where, how, why, should, could and would are answered through the developed theory. This supports the work of Whetten (1989), who outlines three essential elements of theory building as: 'what, how and why?'. The 'what' factors are the constructs that make up the theory, 'how' involves the linkages between the constructs, and 'why' is the search for answers for the existing relationships between constructs which could be due to social or economic factors. These three elements together make the fourth part of good theory; 'description and explanation'. Another element of theory building to add to this is the consideration of the context in which the theory is applied (Whetten, 1989). This is because the impact of time may affect the theory and, if so, this should be outlined when describing any limitations with regard to the application of the theory.

The importance of theory building is highlighted by Wacker (1998), who states that: 'It provides a framework for analysis, facilitates the efficient development of the field, and is needed for the applicability to practical real world problems'. Allan (2003) defines the use of grounded theory as investigating actualities in the real world. This is supported by Selden (2005), who states that a theory should be practically useful. Theory building explores new areas as well as building on existing theory.

What makes a robust theory is the similarities across concepts across different domains (Wacker, 1998). Wacker (1998) cites Van de Ven (1989), who states that: 'good theory is practical precisely because it advances knowledge in a scientific discipline, guides research toward crucial questions, and enlightens the profession of management'. This point is made by Whetten (1989), who outlines the importance of new knowledge that challenges the old when developing a theory. Finally, the development of a good theory should discuss how it is to be used in reality (Wacker, 1998).

The derived theory

Different methods may be used for generation of the theory (Glaser and Strauss, 1967). 'Theory is in a continual process of reformulation and development as it is applied' (Glaser and Strauss, 1967). This is supported by Denzin and Lincoln (2000), who suggest that theory is limited in time where theories will eventually become outdated. Therefore, this must be taken into consideration as the theory is built on and new data is introduced. Future work in grounded theory research studies may involve enhancing the theory to determine what new concepts can be added to the theory and to generate a formal theory which is more generic in scope from the substantive one developed.

The theory should outline its contribution to knowledge, further work recommended to increase the scope of the theory and the benefit to those in the field. Generally, the theory will permit a clearer understanding of the substantive area investigated. This is supported by Strauss and Corbin (1998), who state that: 'a theory explains and predicts events, thereby providing guides to action'. Supporting this, Douglas (2003) suggests that: 'the explanatory power of the grounded theorist is to develop predictive ability'. Glaser and Strauss (1967) suggest that the practitioner will have more control and a better understanding of situations from the explanation derived through theory. Denzin and Lincoln (2000) highlight the point that not every grounded theory may be used for prediction or have a practical application, where some theories may be used purely as a source of understanding.

Summary

This chapter has summarised the debate on grounded theory which began with the seminal work of Glaser and Strauss in 1967. That Glaser and Strauss subsequently disagreed on the methods for undertaking grounded theory is indicative of the opportunities which face the researcher commencing a research project. The criticism that grounded theory is not scientific reflects the traditional view of basic or pure science which begins with an observation of the natural world and attempts to explain a precise hypothesis through a repeatable, often laboratory-based, experiment. The result of the endeavour is to prove or disprove the hypothesis. Basic or pure science is the world of the positivist. Grounded theory is an inductive research methodology which involves a systematic process of gathering and analysing a finite set of data to evolve a theory based upon the data. The use of grounded theory lays a huge responsibility on the researcher. This chapter has outlined the various concepts and characteristics of good theory building together with factors which the researcher using grounded theory should keep to the fore.

References

Allan, G. (2003) A critique of using grounded theory as a research method, *Electronic Journal of Business Research Methods*, 2(1), 1–10.

Backman, K. and Kyngas, H.A. (1999) Challenges of the grounded theory approach to a novice researcher, *Nursing and Health Sciences*, 1, 147–153.

Burca, S. de and McLoughlin, D. (1996) *The Grounded Theory Alternative in Business Network Research*, City University Business School, Dublin.

Burrell, G. and Morgan, G. (1979) *Sociological Paradigms and Organisational Analysis*, Heinemann, London.

Charmaz, K. (1990) "Discovering" chronic illness: Using grounded theory, *Social Science and Medicine*, 30, 1161–1172.

Coffey, A. and Atkinson, P. (1996) *Making Sense of Qualitative Data: Complimentary Research Strategies*, Sage, London.

Crabtree, B.F. and Miller, W.L. (1999) *Doing Qualitative Research* (2nd Edition), Newbury Park, CA: Sage.

Creswell, J.W. (1994) *Research Design, Qualitative and Quantitative Approaches*, Sage, Thousand Oaks, CA.

Cronholm, S. (2002) Grounded theory in use – a review of experiences, in Remenyi, D. (ed.) *European Conference on Research Methodology for Business and Management Studies*, Reading University, UK, April 29–30, 2002.

Cutcliffe, J.R. (2000) Methodological issues in grounded theory, *Journal of Advanced Nursing*, 31(6), 1476–1484.

Dainty, A.R.J., Bagihole, B.M. and Neale, R.H. (2000) Computer aided analysis of qualitative data in construction management research, *Building Research and Information*, 28(4), 226–233.

Denzin, N.K. and Lincoln, Y.S. (2000) (eds) *Handbook of Qualitative Research* (2nd Edition), Sage, Thousand Oaks, CA.

Douglas, D. (2003) Inductive theory generation: A grounded theory approach to business inquiry, *Electronic Journal of Business Research Methods*, 2(1).

Edwards, M., McConnell, R. and Thorn, K. (2002) Constructing reality through grounded theory: A sport management study, in Remenyi, D. (ed.) *European Conference on Research Methodology for Business and Management Studies*, Reading University, UK, April 29–30, 2002.

Eisenhardt, K. (1989) Building theories from case study research, *Academy of Management Review*, 14(4), 532–550.

Esteves, J., Ramos, I. and Carvalho, J. (2002) Use of grounded theory in information systems area: An exploratory analysis, in Remenyi, D. (ed.) *European Conference on Research Methodology for Business and Management Studies*, Reading University, UK, April 29–30, 2002.

Glaser, B. (1978) *Theoretical Sensitivity*, Sociology Press, Mill Valley, CA.

Glaser, B. (1992) *Emergence v Forcing Basics of Grounded Theory Analysis*, Sociology Press, Mill Valley, CA.

Glaser, B.G. and Strauss, A.L. (1967) *The Discovery of Grounded Theory: Strategies for Qualitative Research*, Aldine, Chicago.

Goulding, C. (2005) Grounded theory, ethnography and phenomenology, a comparative analysis of three qualitative strategies for marketing research, *European Journal of Marketing*, 39(3/4), 294–308.

Haig, B.D. (1995) Grounded theory as scientific method, *Philosophy of Education*, http://www.ed.uiuc.edu/EPS/PES-yearbook/95_docs/haig.html

Heath, H. and Cowley, S. (2004) Developing a grounded theory approach: A comparison of Glaser and Strauss, *International Journal of Nursing Studies*, 41(2), 141–150.

Hunter, K., Hari, S., Egbu, C. and Kelly, J. (2005) Grounded theory: Its diversification and application through two examples from research studies on knowledge and value management, *The Electronic Journal of Business Research Methodology*, 3(1), 57–68, www.ejbrm.com.

Kelly, J. (2004) A proposition for a construction research taxonomy, *ARCOM (Association of Researchers in Construction Management)*, Heriot Watt University, Edinburgh, Vol. 2, 1175–1186.

Kinach, B.M. (1995) Grounded theory as scientific method: Haig-inspired reflections on educational research methodology, *Philosophy of Education*, http://www.ed.uiuc.edu/EPS/PES-yearbook/95_docs/kinach.html

Miles, M.B. and Huberman, A.M. (1984) *Qualitative Data Analysis: A Sourcebook of New Methods*, Sage, Beverley Hills.

Miles, M.B. and Huberman, A.M. (1994) *Qualitative Data Analysis* (2nd Edition), Sage, Thousand Oaks, CA.

Mintzberg, H. (1979) *The Structuring of Organisations: A Synthesis of the Research*, Prentice-Hall, Englewood Cliffs, NJ.

Pfeffer, J. (1982) *Organizations and Organizational Theory*, Marshfield, MA: Pitman.

Poole, M.S. and Van de Ven, A.H. (1989) Using paradox to build management and organisation theories, *The Academy of Management Review*, 14(4), 562–578.

Robrecht, L.C. (1995) Grounded theory: Evolving methods. *Qualitative Health Research*, 5(2), 169–177.

Scott, M., Davidson, J. and Edwards, H. (2002) Application of template analysis in information systems research: A technique for novice researchers, in Remenyi, D. (ed.) *European Conference on Research Methodology for Business and Management Studies*, Reading University, UK, April 29–30, 2002.

Seidel, J.V. and Kelle, U. (1995) Different functions of coding in the analysis of textual data, in Kelle, U., Prein, G. and Bird, K. (eds) *Computer-aided Qualitative Data Analysis: Theory, Methods and Practice*, Sage, London, pp. 52–61.

Selden, L. (2005) On grounded theory – with some malice, *Journal of Documentation*, 61(1), 114–129.

Spiggle, S. (1994) Analysis and interpretation of qualitative data in consumer research. *Journal of Consumer Research: An Interdisciplinary Quarterly*, 21(3), 491–503.

Star, S.L. (1997) Grounded classification: Grounded theory and faceted classification, *Information Systems and Qualitative Research*, IFIPS WG 8.2, Philadelphia, May 1997.

Strauss, A. (1987) *Qualitative Research for Social Scientists*, Cambridge University Press, Cambridge.

Strauss, A. and Corbin, J. (1990) *Basics of Qualitative Research: Grounded Theory Procedures and Techniques*, Sage, Newbury Park, London.

Strauss, A. and Corbin, J. (1998) *Basics of Qualitative Research, Techniques and Procedures for Developing Grounded Theory* (2nd Edition), Sage, London.

Tesch, R. (1990) *Qualitative Research: Analysis Types and Software Tools*, Falmer Press, London.

Van de Ven, A.H. (1989) Nothing is quite so practical as a good theory. *The Academy of Management Review*, 14(4), 486–489.

Wacker, J.G. (1998) A definition of theory: Research guidelines for different theory-building research methods in operations management, *Journal of Operations Management*, 16, 361–385.

Whetten, D.A. (1989) What constitutes a theoretical contribution? *The Academy of Management Review*, 14(4), 490–495.

Woodhead, R. and Downs, C. (2001) *Value Management: Improving Capabilities*, Thomas Telford Publishing, London.

Yin, R.K. (1994) *Case Study Research, Designs and Methods* (2nd Edition), Sage, London.

Chapter Nine

Case study research

David Proverbs and Rod Gameson

Introduction

Case study research appears to be highly relevant to an industry that is project driven and made up of many different types of organisations and businesses. However, application of the approach within the construction management research community is seemingly at a relatively low level. Perhaps this is due to the lack of guidance on the application of case study research techniques within the specific context of the built environment. Indeed, there exists some confusion as to exactly what merits 'case study' research and often students and academic staff alike have been known to misinterpret the term.

This chapter provides an introduction and guide towards implementing case study research. Use is made of an example, that is a case study research project, presented towards the end of the chapter, to help demonstrate the application of the technique through the various phases of the research, incorporating design and selection through to analysis and writing-up. The emphasis is towards providing a well-grounded and yet practical guide to adopting the case study technique within construction research. This chapter commences with a generic introduction to the case study strategy before going into more specific details of the design and execution of such an investigation. A checklist of common issues to consider when contemplating the use of a case study approach is included, at the end of this chapter, to help guide researchers.

Case study research: An overview

Case study is a strategy used to research an experimental theory or topic using set procedures, often comprising several different combinations of data collection such as interviews and documentary evidence, where the emphasis is towards investigating a phenomenon within a context (Fellows and Liu, 2003). Case study research often adopts the use of triangulation in using three sources of evidence methods (Yin, 2003a). The main advantage of this is that it allows the researcher to evaluate three different sources of information to test a particular concept or theory on the basis that a consensus of the findings will yield more robust results. As such, it commonly uses a certain amount of quantitative data to reinforce the qualitative primary data (Mangen, 1999).

According to Remenyi *et al.* (2002) the following characteristics of a case study should prevail:

- It is a story.
- It draws on multiple sources of evidence.
- Its evidence needs to be based on triangulation of these sources of evidence.

- It seeks to provide meaning in context.
- It shows both an in-depth understanding of the central issue(s) being explored and a broad understanding of related issues and context.
- It has a clear-cut focus on either an organisation, a situation or a context.
- It must be reasonably bounded. It should not stretch over too wide a canvas, either temporal or spatial.

Designing case studies

In determining how to undertake a case study investigation, the researcher should give consideration to the following:

- Time available to carry out the investigation.

This will largely determine whether a longitudinal or a cross-sectional study is possible. With longitudinal case studies the focus is towards investigating a subject to identify change or development over a period of time; for example, monitoring health and safety levels on a project from commencement of construction work to completion. Cross-sectional studies, which aim to capture the situation at that moment in time, that is a snapshot, are therefore less time demanding and can generally be used to obtain findings more quickly.

- Availability of documentary information.

A key component of case study research focuses upon using existing information that emanates from the unit of study, that is from a construction project or construction company for example. This may be in the form of project information such as drawings, specifications, contractual documentation, correspondence including letters to and from the architect or client, minutes of relevant site or project meetings, policy documents belonging to the company such as Health and Safety and many others. Access to this kind of information is therefore important and may be restricted due to confidential matters and/or sensitivity of the topic under investigation.

- Access to persons involved (e.g., for interviewing purposes).

Interviews are often considered as one of the most important sources of case study information (Yin, 2003a). These should be designed to target people directly involved with the case(s) concerned and allow a detailed insight into the subject(s) to be developed.

- Aim or thrust of the investigation.

The subject of investigation will determine the focus of the data collection techniques and their significance to the study. The aim and objectives of the study will often dictate the nature of the investigation and, for example, whether a cross-sectional or longitudinal study might be preferred.

- Number of cases.

Whether to focus on one particular case or a number of cases is a complex issue. There are advantages in being able to compare and contrast findings from one case to a similar or related case. However, this has to be balanced by the distribution of resources across two or more cases, which can affect the depth of the investigation and to some extent the validity of the research findings. This also raises the issue of how to select the cases and on what basis should potential cases be considered for adoption in the study. Should one go for a case that is highly typical and therefore the findings will be of relevance to many other similar projects? Alternatively, a one-off case or project that has particular characteristics may offer the opportunity to discover novel findings not previously reported, but potentially be of limited relevance and therefore value to other more typical examples.

This naturally leads us into dealing with the issue of how to identify and select cases for the investigation.

Identifying and selecting the case(s)

In identifying and selecting the case(s) for investigation, a key issue concerns an understanding of the unit of analysis. In a built environment context, this may be a construction project, a company or organisation, or an individual or group of individuals, such as project managers, architects and so forth. The choice of unit will largely be determined by the aim and objectives of the study. For example, if one were studying maintenance management in a local authority, then the unit of assessment would be the local authority. A single case study would focus on investigating a particular local authority chosen for specific reason involving the detailed exploration and scrutiny of that particular organisation. A multiple case approach would involve two or more local authorities, perhaps chosen to demonstrate distinct characteristics or similarities/differences.

This leads on to how to go about determining the number of cases, or local authorities in this example, that will best help to address the aim of the research. Yin (2003a) provides five reasons for choosing to adopt a single case study as follows:

- The critical case for use in testing a key theory or concept.
- The extreme or unique case or highly unusual case.
- The representative or typical case to capture everyday occurrences.
- The revelatory case providing the opportunity to observe a previously unseen phenomenon.
- The longitudinal case involving the study of the same case at two or more different points in time.

Whilst these represent logical reasons for adopting a single case, it must be said that the results of such investigations will to some extent always be treated with some degree of circumspection due to the fact that they are drawn from one case and no one can be sure as to how they apply to other cases. Hence, when undertaking single case study projects, careful selection and investigation of the chosen case is very important to ensure that the case is meeting the criteria determined by the objectives of the study.

With multiple case studies, even those containing just two cases, the results will always be more compelling, assuming that they are in support of each other and therefore easier to defend. However, extensive additional resources are likely to be needed to undertake the study, possibly beyond the realms of one researcher as is the case for most postgraduate work and again particular emphasis should be given to the selection of cases. Perhaps the overriding consideration should be to maximise what can be learnt from the case(s) (Stake, 1995).

Collecting the information

All evidence will be of interest to the case study researcher, albeit it will vary in relevance and reliability. As such, the design of case study investigations should incorporate different kinds of evidence. Gillham (2000) and Yin (2003a) have both generally grouped these sources into the following six categories:

- Documents.
- Archival records.
- Interviews.
- 'Detached' or direct observations.
- Participant observation.
- Physical artefacts.

A brief description of these categories is now presented.

Documents

These can be letters or correspondence, minutes of meetings; drawings and contractual documentation; bills of quantities; daily, weekly or monthly reports; records of health and safety inspections and so on. Such documents are important, as they can be useful in helping to corroborate evidence from other sources and in obtaining some basic factual information about the case at hand. That is, this documented evidence will be useful in laying the foundations for the study.

Archival records

This is more likely to be information focusing upon the past of the company, or organisation, or project under study and may well be in the form of computer files and records, possibly located at the head office. How relevant such information would be will be determined by the aim of the study. It should however be borne in mind that, as for documents, this evidence was produced for a specific purpose that is likely to be different to the purpose of the case study. As such, it is important that this evidence is treated objectively with due care and caution.

Interviews

Interviews represent a very important aspect of case study research and are used to fully understand someone's impressions or experiences, or to learn more about their answers to questions. They will often involve open-ended questions and a degree of flexibility in order to probe and delve into issues as they arise. In such cases, interviewing of key persons representing different perspectives on the case, such as contractors' and architects' perspectives, can be employed to develop a fuller picture of the situation. Alternatively, a structured approach may be used when more quantitative information or data is needed – in effect surveying the views of particular employees or parties to the case. This may be compared to surveying these individuals and as such treated very much the same as a questionnaire survey. The main advantages provided by interviewing include obtaining a full range and depth of information and in developing a relationship with the interviewee that might be of use later in the study.

Detached or direct observations

Detached or direct observations involve 'watching from the outside' and, like all forms of observation, they represent a valid form of data collection in that they record what people actually do, that is rather than what they say they do. This form of observation involves the researcher observing and recording what they see and hear. The intention is to minimise the effect of the researcher's presence in order to capture an unbiased and accurate reflection of people's actions and practices. This might involve observing construction operatives in a similar way to that employed in work-study techniques (Harris and McCaffer, 2001), or observing the use of a new type of construction technology. Typically, information observed is recorded onto a pre-designed form or template and then analysed, often statistically, at a later time.

Participant observation

Where detached observation might be considered as the quantitative form, participant observation might be considered the qualitative version of the technique. Here instead of observing from a distance, the researcher is actually immersed in the situation or unit of analysis, for example, as an employee engaged on a construction project under investigation.

The underlying concern with the use of observations is the propensity for researcher bias in terms of what and who is observed and how this is perceived. It is important therefore that clear criteria and guidance is followed to minimise this effect. The influence of the presence of the researcher/observer should not be ignored but instead acknowledged when interpreting the findings and drawing conclusions. Conducting observations of any sort is extremely time-consuming and give the possible limitations, their value in the overall case study design should be fully considered. They may, however, be used at various stages of a research study from early exploration of the subject through to representing the main method of data collection as in studies on productivity.

Physical artefacts

This form of data collection has been used extensively in anthropological research. Artefacts can be things that are made and produced ranging from completed buildings to individual building components or products and incorporating architectural models and tools or instruments. In construction research the scope for this is endless. Artefacts may be observed as part of a visit to site or collected for example from a factory and scrutinised later in laboratory conditions.

Analysing the information

Yin (2003a) recommends developing an analytical strategy towards preparing and conducting case study analysis and that this should be in place well before any information is collected. The most obvious approach is to focus on using the original objectives or research questions of the study to help determine and guide the case study analysis – what Yin describes as following the theoretical propositions. An alternative approach might be to consider rival explanations or theories, which can be investigated through the data collection techniques employed. This strategy can be linked back to the original hypothesis developed for the study and has similarities with the theoretical proposition strategy. A third approach identified by Yin is towards developing a case description, which sets about developing a descriptive framework for the case study.

With regard to the utilisation of specific techniques for analysing the data gathered for case study research, this will usually depend upon the type and volume of information gathered, and may include both qualitative and quantitative data. There are now numerous software applications for analysing both forms of data and as coverage of these is outside the scope of this chapter readers are directed to Lewins and Silver (2007), Pallant (2003) and Chapter 12 in this book.

Writing up

The production of a case study research report is often demanding, as this will involve the bringing together of different kinds of evidence that must be woven together to form a coherent narrative. The same generic advice pertains to writing for your audience, writing clear sentences and structuring the writing around appropriate paragraphs. The structure of the report may often conform to the conventional report or thesis format, commencing with the literature review before describing the research methodology, data analysis and findings, conclusions and recommendations. Alternatively, the structure might follow the chronological phases, for example, reporting a longitudinal study where the sequence of chapters will mirror the findings from the case study as conducted. Other forms of structure worthy of consideration include comparative, theory-building, suspense and unsequenced structures (Yin, 2003a).

Example

There now follows an example of the application of case study research in construction on the subject of project team dynamics. This is provided as a means of demonstrating how the case study approach can or might be applied in a construction management context. The example should not be considered as an exemplar but rather as a guide to how the various stages of the approach can be tackled and how these influence the nature of the findings and conclusions able to be drawn from the study.

Case study – project team dynamics: A scoping study

The construction industry offers its services to customers. Change in the industry has been significantly driven by large client organisations demanding a better service from construction industry participants. A key factor in the achievement of successful project outcomes is the nature of the relationship between members of project teams, including clients, which needs to be established as early as possible in a project's life cycle. This research aims to investigate project team dynamics from the standpoint of 'people and process' issues. The impact of the growth of Information and Communication Technology (ICT) on teams was reviewed, specifically in terms of how its implementation affects project teams, and the individuals within them. The research method involved a literature review which identified key factors relevant to project and Virtual Team (VT) environments, and then developed a case study methodology for conducting interviews with participants in a construction supply chain. After this, data were analysed qualitatively and conclusions were drawn by comparing and contrasting theory with practice.

Having reviewed literature relevant to both project team dynamics and virtual teams, 24 key issues were identified, including: relationships between team members, skills development, organisational design and structure, project success criteria and recognition of the need to employ different skills. The review of literature, in both project team and virtual team domains, supported the contention that both people and process issues are important when considering project team dynamics.

The overriding theme, emerging from the investigation, was that of 'work design', to enable people to work in project teams, utilising virtual environments, efficiently and effectively. To enable a test to be conducted, a model was developed encompassing the key issues from literature.

An existing model was used to develop a test. Parker *et al.* (2001) propose a work design model containing five categories of variables: antecedents, work characteristics, outcomes, mechanisms linking work characteristics to outcomes and contingencies affecting the link between work characteristics and outcomes. This model is the result of a comprehensive review of literature based upon a premise that, 'developments in work design theory have not kept pace with changes in the organizational landscape'. This includes consideration of changes to project teams and the growing influences of new technologies upon the work of such teams.

Given the scoping nature of this study, and the research objective to investigate project teams, data was collected by interviewing participants in a recently completed construction project. This allowed for comparisons to be made between project team members to produce a detailed 'case study' formed from multiple sources of evidence (Yin, 2003a, p. 3). As Jankowicz (1995, p. 172) states, a case study, '… explore[s] issues both in the present and in the past, as they affect a relatively complete organizational unit …'.

The case study project

The case study project was a recently completed large Government facility with a contract value of £40 million and a construction duration of approximately 2 years. The client was a public sector government body. The professional design team were all employed by the client organisation, but had to tender competitively to work on the project. The project was procured using a traditional contract with a contractor tendering on a full design. This was a departure from previous procurement practice, on similar projects, where a design and construct form of procurement had been used involving the same contractor. Once the contractor was appointed, a partnering relationship was established between all of the project participants. One of the contractual requirements was that the main contractor established a 'web portal' to be used by all parties working on the project.

As this was a pilot study, with limited time and resources available, a major aim of this research was to scope the field of investigation before embarking upon a larger, and more detailed, research project. As Yin (2003a) suggests, pilot enquiry can be used to improve conceptualisation of the research domain.

Data collection

Data were collected by conducting semi-structured face-to-face interviews with eight members of the project team, who worked on the case study project. They were:

- Senior project coordinator/project architect.
- Client representative.
- Project manager.
- Senior mechanical engineer.
- Principal electronics engineer.
- Senior quantity surveyor.
- Main contractor.
- Subcontractor.

All interviewees consented to take part in the research project. The objective of this approach was to gain an insight into different views of all parties involved in a construction project.

The Parker *et al.* (2001) model was used to elicit views from participants in the construction process, and to obtain opinions on key factors influencing team dynamics, which in turn influences team performance. The model contains all of the key factors identified from the literature reviewed in this investigation. It was translated into a set of interview questions. Questions were grouped under the main categories in the Parker *et al.* (2001) model. The main sections of the interview schedule were:

- **Section 1: Background Information**. Collected general information on the interviewee's organisation [5 questions].
- **Section 2: Project Antecedents**. Collected data relating to external organizational, internal organisational and individual factors [17 questions].
- **Section 3: Work Characteristics on Case Study Project**. Collected data relating to groups (e.g., teams) and individuals, and the interaction between groups and individuals [14 questions].
- **Section 4: Mechanisms and Contingencies Linking Work Characteristics with Project Outcomes (Intermediary Outcomes)**. Collected data on factors linking and influencing work characteristics and project outcomes. This section combines categories 4 and 5 in the Parker *et al.* (2001) model [10 questions].
- **Section 5: Project Outcomes**. Collected data relating to project organisational and individual factors [10 questions].

Each question had two dimensions, as illustrated by the following sample question (Table 9.1):

Q13. Has your organisation introduced any new communication methods and/or supporting technology during your involvement in this project?

For 'Project case study experiences' the respondent was being asked to recount their experiences of the case study project. This was to enable a 'picture' of experiences of the participants in the case study to be painted.

For 'Views on future changes/improvements' the respondent was being asked to 'reflect' upon their experiences of the case study project, and elucidate how they thought changes and improvements could be made. This dimension was considered important as literature highlighted the usefulness of developing 'learning organizations' (Senge, 1992), generating reflective learning environments (Love *et al.*, 2002, p. 5) and monitoring and reviewing project performance (Hormozi *et al.*, 2000, p. 48).

Consideration of ethical issues is particularly important when conducting interviews because, according to Easterby-Smith *et al.* (1991, p. 82), '… of the potential freedom within the interaction for exchanging information and interpretations'. Therefore, the research methodology proposed here was submitted to the researchers' 'Human Research Ethics Committee' for approval, with approval being obtained before any data was collected.

Other items of data relating to the case study project were available to the researchers; for example project records (e.g., drawings, specifications, etc.). Yin (2003b)

Project case study experiences	Views on future changes/improvements

Table 9.1   Questionnaire Q13.

argues that multiple sources of evidence can improve findings, interpretations and conclusions, by applying concepts such as triangulation. However, given that the focus of this research was on work design, and human and technology interaction in particular, it was decided to concentrate on a series of interviews to elicit the most appropriate data.

Data processing/analysis

The approach adopted for the processing and analysis of data, and the write-up of the study can be explained using the analogy of a series of SIEVEs being used to sort and separate a sample of soil. At specific intervals sub-sets of the overall sample are collected and analysed, ranging from large to small particles. Having completed the analysis, an overall profile of the soil sample is produced.

All eight interviews were tape recorded and then verbatim transcripts were word processed [SIEVE 1]. To assist with and expedite the analysis of the interview data a specialised computer software package was used. The 'Ethnograph' software is, '… a collection of procedures designed to enhance and facilitate the process of qualitative data analysis … the process of noticing, collecting and thinking about interesting things (Seidel, 1998, p. 3).

Interview transcripts were formatted and input to the software [SIEVE 2]. Unique codes were then assigned to the answers to each interview question to enable searching of the database created within the software [SIEVE 3]. Once coded, all eight interview data files could be searched by 'code word' to allow answers [segments] provided by each interviewee to be compiled together for a qualitative analysis to be conducted looking for similarities and differences in responses [SIEVE 4].

Writing up

A qualitative content analysis of the sorted coded segments of the interview data was conducted. Results of the analysis of each question were first summarised in a table and then they were discussed, with verbatim quotations from interviews used to support the discussion, where appropriate [SIEVE 5]. An example of the tabular format summarising interviewees' responses to one question [Q21] is shown below (see Table 9.2):

Section 3:   Work Characteristics on Case Study Project

(A) *Group level*

Q21. If you communicated electronically with other teams working on the project, what factors influenced the efficacy of this communication?

Project Case Study Experiences	Views on Future Changes/Improvements
• Unclear meaning when using emails, requiring verification • Need access for all to web portal system/overcome access problems. • Timeliness of responses; provision of sufficient information.	• Clarity of communication • Improved technology (e.g., utilising video and voice) with infrastructure support. • More familiarisation with ICT systems • Personal preferences: electronic versus face to face/telephone

Table 9.2   Questionnaire Q21.

Finally, the 24 key issues, identified from the literature, were reviewed in the context of the results of the case study [SIEVE 6]. This approach was adopted to allow for an easy comparison to be made between the key issues identified from literature and the two dimensions of data collected from the case study: participants' experiences of the case study project, and their reflections on how changes and improvements could be achieved in the future.

Conclusions

The theoretical foundation of the research, having reviewed literature, proffered two propositions. The first was that the use of ICT could be seen as something which could improve the nature of project team performance by reducing more mundane tasks (such as record-keeping, improving communication, etc.) to allow team members to concentrate energies on being more creative and innovative. The second was a contrary view that the use of ICT, without a strategy to ensure that project team members have the necessary knowledge, understanding and skills to utilise the technology to its potential, could be counterproductive.

For the first proposition, both the literature and findings of the case study supported the case for ICT leading to improvements in the performance of people and projects. The use of systems, such as the web portal utilised in the case study project, did contribute to a perceived successful project outcome. ICT was seen as facilitating timely solutions, with the potential for further development to: incorporate more comprehensive data storage (e.g., drawings and project documentation such as minutes of meetings) and the generation of 'action plans', possibly based upon a database of experiences and 'lessons' learned from previous projects. Such initiatives would promote a culture of business development encompassing organisational learning and continuous improvement.

Evidence to support the second proposition was also present. With regard to 'people' issues the need for training, and the acquisition and continuing development of appropriate knowledge and skills, was seen as crucial. Ensuring that all project participants 'engaged' with systems, such as the web portal used in the case study, was seen as essential to ensure they are used effectively and efficiently. Therefore, support is required in terms of skills development, where appropriate. In addition, 'technology' issues also need addressing. There is a need to ensure that ICT systems are both reliable and integrated (e.g., compatible software platforms are used by all team members) and that developing technologies, such as video cameras, are utilised.

This research concluded that, given the increasing impact and utilisation of ICT, there was a need to investigate further how project 'success' is achieved, particularly in relation to developing strategies for the following three areas:

- 'People' (e.g., knowledge, skills development and training).
- 'Process' (e.g., project performance and outcomes).
- 'Technology' (e.g., ICT systems: hardware and software).

Conclusions

This chapter has presented an introduction and overview of the case study approach. Key towards the approach is investigating a subject in context, rather than at a distance or in some artificial environment. Some important considerations when designing case studies include available time, availability of information, access to persons, aim(s) of the investigation and the number of cases to be considered. While

- Have quantitative approaches been thoroughly considered and eliminated for sound reasons?
- Have alternative qualitative approaches to case study (e.g., action research, focus groups, interviews) been thoroughly considered and eliminated for sound reasons?
- Has the decision to use a single or multiple case approach been made?
- Is access available to the required case study data (e.g., documents, people, processes, technology)?
- Is ethical approval required to allow collection of data relating to the case(s)?
- Do you have the necessary skills to conduct the data collection and analysis of the case(s)?
- If using a single case approach does this represent the critical case, the extreme or unique case, the representative or typical case, the revelatory case and/or the longitudinal case?
- If a multiple case approach is adopted, what is the rationale for opting for two or more cases?
- Have you identified three data collection techniques (e.g., document review, interview, observation) that will allow triangulation of the findings?
- Do you intend to pursue a theoretical, rival theoretical or descriptive perspective in analysing the information gathered?
- How will qualitative and/or quantitative data be analysed?
- Will you utilise conventional, comparative, theory-building, suspense or unsequenced structure for writing up the case report?

Table 9.3   Case study checklist – key questions to consider.

multiple cases are to be preferred there may be much benefit from a single case where this can be identified as, for example, the critical, typical or revelatory case.

Collecting the information is likely to comprise a combination of research techniques including interviews, review of documents and observations, where the intention is to achieve triangulation and convergence of findings from different sources. Once the information has been collected, analysis is likely to involve a combination of qualitative and quantitative techniques, where use of appropriate analytical software will help speed up the process. Writing up may follow the conventional format of research projects or, alternatively, may present a story of the chronological development of the investigation.

The authors opine that there remains considerable scope for further application of the case study technique in studying, capturing and disseminating the innovations and novel solutions adopted on construction projects and/or within construction organisations (see Table 9.3). While the case study approach presents some unique challenges to the researcher, the detailed study of a case can bring about significant improvements and changes in the field of investigation.

Acknowledgement

The example case study presented in this chapter was part of a research project funded by the Cooperative Research Centre for Construction Innovation (CRC-CI) [www.construction-innovation.info], part of the Australian Government's CRC programme.

Notes

1 For more information on key issues to consider when conducting case study research see the 'Exercises' at the end of each chapter in the book by Yin (2003a).
2 Examples of reports of case studies that have been conducted across a wide range of disciplines are available from the 'European Case Clearing House (ECCH)': www.ecch.com

References

Easterby-Smith, M., Thorpe, R. and Lowe, A. (1991) Management Research: An Introduction, Sage, London.

Fellows, R. and Liu, A. (2003) Research Methods for Construction (2nd Edition), Blackwell Publishing, Oxford.

Gillham, B. (2000) Case Study Research Methods, Real World Research, London.

Harris, F.C. and McCaffer, R. (2001) Modern Construction Management (5th Edition), Blackwell Science, Oxford.

Hormozi, A.M., McMinn, R. and Nzeogwu, O. (2000) The project life cycle: The termination phase, SAM Advanced Management Journal, Winter, 45–51.

Jankowicz, A.D. (1995) Business Research Projects (2nd Edition), Chapman & Hall, London.

Lewins, A. and Silver, C. (2007) Using Software in Qualitative Research: A Step by Step Guide, Sage, London.

Love, P.E.D., Irani, Z., Cheng, E. and Li, H. (2002) A model for supporting inter-organizational relations in the supply chain, Engineering, Construction and Architectural Management, 9(1), 2–15.

Mangen, S. (1999) Qualitative research methods in cross national settings. International Journal of Social Research Methodology, 2(2), 109–124.

Pallant, J. (2003) SPSS Survival Manual: A Step by Step Guide to Data Analysis using SPSS for Windows (2nd Edition), Open University Press, Maidenhead.

Parker, S.K., Wall, T.D. and Cordery, J.L. (2001) Future work design research and practice: Towards an elaborated model of work design, Journal of Occupational and Organizational Psychology, 74, 413–440.

Remenyi, D., Money, A., Price, D. and Bannister, F. (2002) The creation of knowledge through case study research, The Irish Journal of Management, 23(2), 1–17.

Seidel, J. (1998) The Ethnograph v5.0: A Users Guide, Qualis Research Associates/Scolari, Sage Publications Software Inc., Thousand Oaks, CA.

Senge, P.M. (1992) The Fifth Discipline: The Art and Practice of the Learning Organization, Random House, Sydney.

Stake, R. (1995) The Art of Case Study Research, Sage, Thousand Oaks, CA.

Yin, R.K. (2003a) Case Study Research: Design and Methods (3rd Edition), Sage, Thousand Oaks, CA.

Yin, R.K. (2003b) Applications of Case Study Research (2nd Edition), Sage, Thousand Oaks, CA.

Chapter Ten

Interviews: A negotiated partnership

Richard Haigh

Introduction

The interview remains a popular method of data gathering by those researching in the built environment disciplines. Its widespread use is likely due to the flexibility afforded by the interview method, from highly structured face-to-face question-naires used in quantitative studies, to open-ended interviews that are used to generate insights and concepts, rather than generalise about them. Despite this popularity, interviews are not an easy option. Interviewing has its own challenges and complexities, and demands its own type of rigour. People, in this context the interviewer and the respondent, are inherently complex, and issues such as completeness, accuracy, tact, precision and confidentiality must all be considered carefully by the researcher. A key feature of some interview forms is the nature of the relationship between the interviewer and the interviewee, who must form a partnership to negotiate a highly detailed and valid set of qualitative data. This places great responsibility on the researcher, who must be aware of the many ways in which he or she can inadvertently influence the interview result, and thereby jeopardise the purpose of the study.

In the space afforded by a single chapter, the author cannot hope to cover interviewing in a comprehensive manner. There is a wide range of books dedicated to interviewing, from Kvale's (1996) succinct accounts of phenomenological and hermeneutic perspectives on interviewing that walk the reader through seven methodological stages of qualitative interview studies, to Gubrium and Holstein's (2001) nine hundred plus page 'encyclopaedia' that provides coverage of the methodological issues surrounding interview practice, including different forms, distinctive respondents, institutional applications, technical matters relating to data processing, and analytical strategies. An annotated bibliography is provided at the end of the chapter as a useful starting point for researchers in the built environment discipline looking to further or deepen their knowledge on research interviews.

This chapter aims to provide an introduction to the variety of forms of research interview that may be used by those researching in the built environment disciplines and to provide practical guidance on how to design and carry out interview research. The chapter begins by examining different forms of research interview, but later focuses specifically upon qualitative research interviews, including their purposes and applications in the built environment discipline. The chapter con-cludes by exploring the challenges associated with planning, conducting and analysing interviews.

The interview method

At the most basic level, interviews are conversations (Kvale, 1996). The interviewer becomes an attentive listener who shapes the process into a familiar and comfortable form of social engagement – a conversation – and the quality of the information obtained is largely dependent on the interviewer's skills and personality (Patton, 1990). The interview method can be used by the quantitative and qualitative researcher alike, and can be used to gather both quantitative and qualitative data. The research interview can take a variety of forms, from highly structured and standardised, to unstructured and free ranging.

The relationship between interviewer and respondent

Interviews are not neutral forms of data gathering, but as Silverman (1997, p. 98) asserts, 'are active interactions between two or more people, leading to negotiated, contextually based results'. The nature of the relationship between the interviewer and the interviewee is an important feature in the interview research method, and it varies according to the theoretical underpinnings of the research. For a quantitative researcher using structured interviews, the interviewee is a 'research subject', in common with someone taking part in an experiment, or completing a questionnaire. A typical model of quantitative research interviewing involves a face-to-face meeting in which a researcher (interviewer) asks an individual (respondent or interviewee) a series of questions with a purpose. The interviewer's job is to enable and facilitate a respondent to focus on a particular subject or theme, and encourage him or her to answer honestly. The interviewer must achieve this without otherwise shaping or influencing the responses, and must therefore remain neutral and objective. Denzin and Lincoln (1998, p. 174) describe this relationship as a 'balanced rapport', casual and friendly, yet decisive and impersonal. For a qualitative researcher, the relationship is part of the process and the interviewee is a participant, rather than a subject. This affords the interviewer a different role, with more relaxed rules and expectations. From a realist epistemological position, interviewees' accounts are treated as providing insight into their organisational lives outside of the interview. In contrast, radical constructionist approaches place greater emphasis on the interview setting, and discourse analysis must be used to understand conventions in speech and the use of the language in a specific context. The interviewer has greater freedom to change the direction of the interview and formulate new questions. In qualitative research interviews, the interviewer's relationship with the respondent is a more open-ended exchange, focused on a particular topic. In this sense, the interviewer and respondent form a partnership to negotiate a highly detailed and ecologically valid set of qualitative data.

Forms of research interview

The interview is widely used to supplement and extend our knowledge about an individual's thoughts, feelings and behaviours, and it can take a variety of forms:

- *Informal, conversational interview*: no predetermined questions are asked, in order to remain as open and adaptable as possible to the interviewee's nature and priorities; during the interview the interviewer 'goes with the flow'.
- *General interview guide approach*: the guide approach is intended to ensure that the same general areas of information are collected from each interviewee; this

provides more focus than the conversational approach, but still allows a degree of freedom and adaptability in getting the information from the interviewee.

- *Standardised, open-ended interview*: the same open-ended questions are asked to all interviewees; this approach facilitates faster interviews that can be more easily analysed and compared.
- *Closed, fixed-response interview*: where all interviewees are asked the same questions and asked to choose answers from among the same set of alternatives. This format is useful for those seeking quantitative data.

The more structured or standardised interview questions are, the more a researcher is able to get quantitative data. The less structured and freer ranging the interview questions, the more qualitative the data becomes. However, as Converse and Schuman note, 'there is no single interview style that fits every occasion or all respondents' (1974, p. 53). Each form has its own uses, strengths and weaknesses, as Table 10.1 illustrates.

Structured interviews

In its simplest form, a structured interview involves one person asking another person a list of predetermined questions about a carefully selected topic. A key feature of structured interview approaches is in the pre-planning of all the questions asked. Structured interviews allow for replication of the interview with others and the possibility to generalise to the population from which the interview sample came. Structured interviews are conducted in various modes: face-to-face, telephone, videophone and the internet. Standardisation helps the reliability of results and conclusions. The interviewer does not deviate from the pre-determined questions or inject any extra remarks into the interview process. However, the interviewer may encourage the interviewee to clarify vague statements or to further elaborate on brief comments. Otherwise, the interviewer attempts to be objective and tries not to influence the interviewee's statements. The interviewer does not share his or her own beliefs and opinions. The structured interview is mostly a 'question and answer' session, whereby the interviewer imposes the concepts, rather than be led by the data.

Unstructured interviews

In the unstructured interview, the concepts emerge as the interviewer explores the area with the respondent. Because the information is not amenable to statistical analysis, this form is often referred to as qualitative interviewing. The interviewer may ask the same sort of questions as in the structured interview, but the style is free-flowing rather than rigid; it is more conversational. The interviewer may adjust the questions according to how the interviewee is responding, or may inject his or her own opinions or ideas in order to stimulate the interviewee's responses. Therefore, the unstructured interview requires much more skill, and is much more complex. Eisenhardt (1989, p. 539) warns:

> This flexibility is not a license to be unsympathetic. Rather, this flexibility is controlled opportunism in which researchers take advantage of the uniqueness of a specific case and the emergence of new themes to improve resultant theory.

Kvale (1996, p. 45) defines qualitative research interviews as, 'attempts to understand the world from the subjects' point of view, to unfold the meaning of people's experiences, to uncover their lived world prior to scientific explanations'. The qualitative research interview seeks to describe and understand the meanings of central themes in the life world of the subjects. Qualitative interviews are particularly useful for getting the story behind a participant's experiences. The interviewer can pursue in-depth information

Form of interview	Strengths/uses	Weaknesses
Structured interview Closed Standardised	All respondents are asked the same questions in the same way. This makes it easy to replicate and closed questions make it easier to obtain quantifiable data. Data are more reliable as the issue is being investigated in a consistent way. Allows generalisation of results/conclusions to the population from which the sample was drawn. It can be used as a powerful form of formative assessment before using a second method (such as observation or in-depth interviewing) to gather a greater depth of information. The researcher is able to obtain, code and interpret data more quickly, easily and efficiently. There is a formal relationship between the researcher and the respondent and unlike postal surveys, the researcher could for example determine why a respondent is unable or unwilling to answer a question.	Restrictive questioning leads to restrictive answers. The quality and usefulness of the information is highly dependent upon the quality of the questions asked. Closed questions are insensitive to participants' need to express themselves. Uncertainty over whether the right questions are being asked, or in the right way for a diverse respondent group to understand them. Relative to a postal survey, it can be time consuming if sample group is very large. A substantial amount of pre-planning is required. The format of structured interview design, particularly if using closed questions, can make it difficult for the researcher to examine complex issues and opinions. There is limited scope for the respondent to answer questions in any detail or depth.
Unstructured interview General guide Informal	Flexible, responsive and sensitive to participants. Relaxed and natural for those taking part. Highly detailed and ecologically valid qualitative data. Allows exploration of subjects where there is a lack of empirical evidence and understanding. Can yield unexpected findings, not previously considered by the researcher.	Difficult to replicate. As a result, an inability to generalise your findings to a wider population. Possible interviewer bias in 'selective' use of leading, and spontaneous questions. It can be very time consuming to transcribe and analyse large quantities of qualitative interview data.

Table 10.1 Forms of interview and their respective strengths and weaknesses.

around the topic. Interviews may also be useful as a follow-up to certain respondents' questionnaire results, allowing the researcher to further investigate their responses.

Despite its apparent flexibility and more relaxed rules, a qualitative interview is very different from everyday conversation. It is a research tool and the interviewer must prepare questions in advance, and later analyse and report results. The interviewer guides the questions and focuses the study. Good interview skills require practice and reflection. Finally, beyond the acquisition of interview skills, interviewing is a philosophy of learning. The interviewer becomes a student and then tries to get people to describe their experiences in their own terms. The results are imposed obligations on both sides. The qualitative researcher's philosophy determines what is important, what is ethical, and the completeness and accuracy of the results.

Structured interviews are frequently criticised as unnatural and restrictive. Informal interviews get deeper. For example, if you want to find out why someone acted in a certain way, ask him or her. One must negotiate an explanation that is consistent and believable. This results in an explanation of the meaning of the action for the people. The interviewer follows up an interview with more questions for clarification or understanding. The key is to establish rapport and trust.

Interviews in the built environment disciplines

As Table 10.2 illustrates, the whole spectrum of interview forms, from qualitative to quantitative, or highly structured to unstructured, can be effectively used to support research in the built environment disciplines. All interview forms have their

---

A study by Pan et al. (2007) on the perspectives of UK housebuilders on the use of offsite modern methods of construction

A survey by Pan et al. (2007) of the top 100 housebuilders in the UK by volume was carried out through a combination of face-to-face and telephone interviews, and a postal questionnaire survey, which yielded an overall response rate of 36 per cent. The interviews were carried out with senior managers that had responsibility for company policy decisions on whether to use off-site modern methods of construction within their developments. An initial survey instrument was developed through a comprehensive literature review of previous studies that investigated the use of off-site modern methods of construction in the past. The instrument comprised a mix of qualitative and quantitative questions, and used Likert scales and other close-ended questions. The interview structure was based on four discrete sections to provide data on the overall views of housebuilders on off-site modern methods of construction applications, the drivers and barriers and their importance or significance, and recommendations for the industry to increase the take-up of off-site modern methods of construction. The instrument was refined through discussions with leading researchers and industrial contacts. The interviews lasted between one and two hours and added rich data to the questionnaire survey of the firms. Pan et al. analysed the data using a combination of Microsoft Excel and QSR NVivo.

A grounded empirical approach for assessing cooperation by Phua and Rowlinson (2004)

Phua and Rowlinson (2004) aimed to quantify and predict the importance of cooperation to project success. They identified a specific difficulty in accurately measuring project success, due to a lack of consensus regarding its meaning and more importantly how it is brought about. On the basis that project success is a complex and multi-faceted concept that cannot be understood using the conventional deductive approach that is based merely on some predetermined, prior selected factors, Phua and Rowlinson employed a three-stage approach to investigate the extent to which cooperation and its correlates, in relation to other determinants, are predictive of project success. The first part of the study involved drawing randomly from a pooled sample, to yield a total of 29 interviews with the most senior identifiable executives, each lasting on an average for an hour. Interviewees were contacted beforehand by telephone and briefed about the nature and objective of the study but specific details about the nature of the questions were not revealed so as to capture spontaneous views and opinions and thereby minimise response bias. In order to reduce method variance, efforts were taken to ensure that the interviews were as minimally structured as possible so that interviewees were free to express their views in an unprompted manner. Although interviews were tape-recorded and supplemented with simultaneous note taking, interviewees were assured complete anonymity and their responses kept completely confidential. The interviews were content analysed to identify all references to project success and the perceived critical project success factors.

Table 10.2 Examples of interview form in the built environment disciplines.

strengths and weaknesses, and preferred usage. The researcher must identify the most appropriate form to support the study's aim and objectives.

The remainder of this chapter will focus on qualitative interviews. In its most structured form, an interview can be very similar to a face-to-face questionnaire survey. Researchers wishing to conduct quantitative interviews therefore face many of the challenges associated with survey and questionnaire research (see Chapter 11).

Purposes of qualitative interviews

Interviews can be used by researchers in the built environment disciplines for a variety of purposes. Qualitative interviews have the potential to generate insights and concepts, and expand our understanding. They can also be used to search for exceptions to the rule by charting extreme cases. Alternatively, research results from other methods can be validated with interviews; such as with members of, construction project team.

Several types of qualitative interviews exist. For example: topical history; life history; oral history; evaluation interview; focus group interview; and cultural interviews. Topical interviews are concerned with the facts and sequence of an event. They focus on a specific subject and target individuals associated with the subject. The interviewer is interested in a reconstruction of the experience and what happened; for instance, why a construction project went over budget or the relationship between a client, the design team and a construction team on a specific construction project. The researcher actively directs questions in pursuit of precise facts. Life histories deal with individual experiences or rites of passage, such as the experience of a female manager in the construction industry. In oral histories, one collects information about a dying lifestyle or art skills. These result in narratives and stories that interpret the past. Evaluation interviews examine new programmes or developments and suggest improvements. Since evaluation deals with incorrect behaviours as well as positive one's, justifications of behaviours result. The result may consist of myths and unresolved tensions (Patton, 1990). Evaluation interviews might be used by a researcher in a construction company or project to review practices and initiate continuous process improvement. In focus group interviews, people meet to share their impressions and changes of thinking or behaviour regarding a product or an institution, such as a type of material, construction technology or professional body. The cultural interview focuses on the norms, values, understandings, and taken-for-granted rules of behaviour of a group or society. This type of interview reports on typical shared activities and their meanings. The style of interview is relaxed and questions flow naturally with no fixed agenda. People are interviewed several times so that emerging themes are pursued later. The interviewer, for example, may ask a site worker or manager to describe a typical day on a construction project. The respondent then relates what is important with examples. The truth of the fact is not as important as how well it illustrates the cultural premises and norms of the site or project team. In the cultural interview, the interviewer is partner and co-constructs the interview and report. The cultural report, besides being the expert's story, is credible because it consists of the words of members of the culture. We assume that people are basically honest and that they share similar views. The researcher can mix types of interviews and approaches.

Planning, conducting and analysing interviews

The power and flexibility afforded by the research interview comes at a price. Planning and preparing for qualitative research interviews, and later gathering and analysing qualitative interview data, are all highly time-consuming activities for the researcher and the respondents. King (2004) describes a four-stage process of constructing and carrying out qualitative research interviews: defining the question, which should focus on how participants describe and make sense of particular element(s) of their lives; creating an interview guide that is a list of topics to be covered in the interview and a list of probes to elicit further details if required; recruiting participants, including sample definition and criteria, and consideration of confidentiality; and, carrying out and analysing the interviews, which addresses the practical issues associated with interviews, such as phrasing, starting and ending, and difficult interviewees. Similarly, O'Leary (2004) presents a step-by-step approach to interview research with phases of planning, preparation of interview schedule and data recording, interview and analysis. O'Leary adds a need to have a pilot interview before beginning a final interview schedule, which includes the need to gather feedback, reflect and where appropriate, modify the interview plan.

In particular, most researchers will find the interviews tiring to carry out due to the level of concentration required. There are three and sometimes four roles that the interviewer must simultaneously fulfil, unless the researcher is fortunate enough to have a colleague to assist. Firstly, the interviewer must question and probe the respondent in order to make the most of the opportunity and collect the 'best' data. Although many questions may be scripted, the qualitative interview affords an opportunity to explore issues not previously considered or that have emerged spontaneously during the interview. Secondly, the interviewer is attempting to listen to the respondent and make sense of what they are saying. Thirdly, the interviewer must also manage the process, including the time and sequence. Finally, in some instances, the interviewer may also elect to take detailed notes for later analysis. This may range from highly structured notes based on a pre-conceived template with a list of common responses or codes, to unstructured note-taking that makes use of a concept map or involves some interpretative analysis. Inevitably such approaches form a preliminary analysis, rather than raw data, and must therefore be managed carefully. An alternative approach is to make use of audio or video recording to collect raw data for later analysis, thereby eliminating or reducing the significance of this fourth and final role.

Regardless of how responses were recorded, the transcription and analysis phases often pose the greatest challenge for researchers using qualitative research interviews. In particular, even a moderate sized study can overload the researcher as a result of the rich volume of data produced. Typically it will take three or four hours to convert the spoken word from a one hour interview into written text. Although time consuming and sometimes tedious, transcription provides the researcher with an opportunity to re-familiarise himself or herself with the data that has been collected. After transcription, the study moves into the analysis phase. In a single chapter it is not possible to cover the wide range of techniques that are available for analysing qualitative research data. As a starting point for further reading, Cassel and Symon (2004) and Miles and Huberman (1994) both cover a range of techniques, from data matrices to cognitive mapping. It is, however, an imperative that analysis is considered prior to the collection of raw data. In common with other methods, the researcher's preferred analysis technique will likely have a significant impact on how the raw data should be collected.

The content versus the process of the interview

The content of the interview is what the interviewee says. This is the easiest component of the interview to study, and tends to be what the novice researcher focuses on.

The most accurate way to record the content of the interview is by using a tape recorder.

The process of the interview is a much more elusive but powerful component of the interview. It involves reading between the lines of what the interviewee says. It involves noticing how he or she talks and behaves during the interview. How the interviewee responds will give the researcher more insights into the content of what he or she says. The researcher's observations of the interview process may confirm, enrich, and sometimes even contradict the content of what the person says.

To explore the interview process, a researcher may consider:

- When does the respondent sound confident or uncertain, confused or clear, convincing or doubtful, rational or illogical, etc.?
- Does the respondent ever contradict himself or herself?
- How do the pieces of what the respondent says fit together?
- At what points does the respondent show enthusiasm and emotion, and what kinds of emotion?
- What is the respondent's body language; when does it change?
- How does the respondent speak: slow or fast, soft or loud, clear or murmur, with simple or elaborate language, etc.?
- Does the respondent's appearance or surroundings (e.g., his or her office) provide any insights?

Because of its power, consideration of the process also requires considerable skill and careful application. It is easy for the researcher to inadvertently influence the interview, or arrive at unsupported assertions.

Influencing the interview

The difficulty of preventing bias in the interview process is an often cited weakness in interview research, which has the potential to undermine the validity of the study's results. The interviewer can inadvertently bias the results, thereby jeopardising the purpose of the study. If the interviewer influences that which is measured, the respondent may tell you what he or she thinks you want to hear, or tell you the opposite of what he or she thinks you want to hear; the important thing is the respondent does not tell you 'the truth'. There are many ways in which the interviewer may inadvertently influence the respondent:

- *Tact*: Did the interviewer protect the respondent's self-esteem and make it easy to express unpopular opinions?
- *Phrasing*: Did the interviewer load the deck by suggesting an answer?
- *Principle of tacit assumption*: Did the interview involve any unspoken communication that was open to misinterpretation?
- *Clarity*: Did the interviewer use long questions, double meanings, double negatives and two part questions, all of which are open to misinterpretation?
- *Completeness*: Did the interviewer record the interview and transcribe the interview exactly as recorded?
- *Bias*: Did the interviewer 'add' to what you observed by presuming or assuming something that was not stated directly by the participant?
- *Accuracy*: Would someone else who had not interviewed the participant be able to get a clear, correct picture of what was discussed by reading the interviewer's notes?

- *Confidentiality*: Did the interviewer ask permission for the interview, and is the participant aware of the purpose and intended audience of the interview?

Reporting interview data

The interview data should be an important part of the researcher's final paper or thesis. The data is not objective as in quantitative research methods. If the interviewee is an expert on some particular topic or possesses some special skill or experience, his or her responses may be 'facts' or 'opinions'. Brief quotes or references to what people said is a waste of the interview. Quotes that are out of context are also insufficient. The researcher's goal is to thoroughly integrate the interview data into the topics and themes of the paper or thesis. Consider these questions:

- Does the interview data support or contradict the researcher's ideas?
- Did what the respondent say support what was uncovered in the literature review? If not, what might this mean?
- Did the respondent support or contradict the other interviewees?
- Did the respondent add new dimensions to what was uncovered in the literature review or to what other interviewees said?
- How did the 'process' of the different interviews compare, and does this reveal any insights concerning the researcher's ideas?

When an interpretative researcher analyses the results from an interview, he or she has the option of using the 'hermeneutic method', whereby he or she examines how all the statements made by the interviewee are inter-related. What are the contradictions and consistencies? What is the 'big picture' of what the interviewee is trying to say – and how does every individual statement from the respondent relate to this big picture? In this sense, the interview is a holistic research method: all the bits of data from the interviewee provide this 'big picture' that transcends any one single bit of data. A good interview is the art and science of exploring the subjective knowledge, opinions and beliefs of an individual. The knowledge, opinions and beliefs of that person are a system. The purpose of the interview is to explore that system and all of its elements.

There are several ways to cite interviewees, use quotes and refer to information from the interviews. Firstly, the researcher can summarise in his or her own words what was said. Secondly, the researcher can use short quotes of phrases or a couple of sentences and embed them into a paragraph. Finally, for longer quotes, the researcher can use a separate indented paragraph. If quoting directly, as in the second and third methods, it is important to identify something important, rather than trivial or obvious.

In the method section of the paper or thesis, it is important to describe who each of the interviewees are, why they were asked to participate, and how they were located. Interviewees who are professionals or 'experts' on a particular topic should be identified by name, profession, where they work, the details of their expertise, and any other information about them that is relevant to the research. Other interviewees should be identified by name, age, occupation, role, or other criteria that are relevant to the specific research project.

In some instances, the researcher will not be able to mention the identities of the interviewees. Permission must always be sought – as part of the informed consent before an interview – to mention a person's name in a paper or thesis. For a person who wishes to remain anonymous it is desirable to include age, marital status, occupation, role, and any other information that is relevant to the research.

False names can be used to improve the narrative. However, it is important not to mention any information that is so specific or unique that it could reveal who the respondent is.

Conclusion

Different forms of research interview can be used to tackle different types of research questions in the built environment disciplines. They are ideally suited to examining topics in which different levels of meaning need to be explored and can be of great use in studying discipline identities in large construction project organisations, where a complex pattern of organisational, group, professional and personal loyalties exist. The flexibility of interview research also places great challenges on the researcher, who must develop a broad range of skills to effectively plan for, collect, transcribe and analyse interview data. Moreover, as one half of the relationship between the interviewer and interviewee, the researcher must work in partnership with the respondent to carefully negotiate a highly detailed and valid set of qualitative data.

References

Cassel, C. and Symon, G. (2004) (eds) *Essential Guide to Qualitative Methods in Organizational Research*, Sage, London.

Converse, J. and Schuman, H. (1974) *Conversations at Random*, Wiley, Ann Arbor, MI.

Denzin, N. and Lincoln, Y. (1998) *Collecting and Interpreting Qualitative Materials*, Sage, Thousand Oaks, CA.

Gubrium, J. and Holstein, J. (2001) (eds) *Handbook of Interview Research*, Sage Publications, London.

King, N. (2004) Using interviews in qualitative research, in Cassell, C. and Symon, G. (eds) *Essential Guide to Qualitative Methods in Organizational Research*, Sage, London, pp. 11–22.

Kvale, S. (1996) *Inter Views: An Introduction to Qualitative Research Interviewing*, Sage, Thousand Oaks, CA.

Miles, M. and Huberman, A. (1994) *Qualitative Data Analysis: A Sourcebook of New Methods*, Sage, Beverley Hills, CA.

O'Leary, Z. (2004) *The Essential Guide to Doing Research*, Sage, London.

Pan, W., Gibb, A.G.F. and Dainty, A.R.J. (2007) Perspectives of UK housebuilders on the use of offsite modern methods of construction, *Construction Management and Economics*, 25(2), 183–194.

Patton, M. (1990) *Qualitative Evaluation and Research Methods* (2nd Edition), Sage, London.

Phua, F.T.T. and Rowlinson, S. (2004) How important is cooperation to construction project success? A grounded empirical quantification, *Engineering, Construction and Architectural Management*, 11(1), 45–54.

Silverman, D. (1997) *Qualitative Research: Theory, Methods and Practice*, Sage, London.

Further reading

The suggestions below are offered as useful starting points for researchers in the built environment discipline to deepen their knowledge of the diverse methodological literature that address interview research:

Cassel, C. and Symon, G. (2004) (eds) *Essential Guide to Qualitative Methods in Organizational Research*, Sage, London. (This provides a balance of theory and practical advice on a range of analysis techniques that can be used with qualitative interview research.)

Gubrium, J. and Holstein, J. (2001) (eds) *Handbook of Interview Research*, Sage Publications, London. (Nine hundred plus page 'encyclopaedia' that provides coverage of the methodological issues surrounding interview practice, including different forms, distinctive respondents, institutional applications, technical matters relating to data processing, and analytical strategies.)

Kvale, S. (1996) *Inter Views: An Introduction to Qualitative Research Interviewing*, Sage, Thousand Oaks, CA. (Succinct accounts of phenomenological and hermeneutic perspectives on interviewing that walk the reader through seven methodological stages of qualitative interview studies.)

Chapter Eleven
Questionnaire design and factor analysis

Mike Hoxley

Introduction

This chapter considers the use of questionnaires as research instruments. Often seen as an easy option, in reality their use requires careful design and pre-planning and invariably the analysis of the data they generate requires a statistical approach to be taken. Their construction, administration, data entry and data analysis, together with the particularly useful statistical technique of factor analysis, are all considered. The author has used survey questionnaires on many occasions, has supervised their use by several research students and this chapter draws heavily upon this practical experience.

Questionnaires are research instruments for use in a survey setting and are intended to *measure* something which may include people's attitudes. The instrument can be administered by post, in face-to-face interview, over the telephone or increasingly by email or directly over the web. If the phenomenon to be measured describes something, for example the voting intentions of the electorate in a political poll, then the survey can be categorised as *descriptive*. On the other hand, if the measurement is looking for association or causality, for example the effect of procurement route on profitability, then the survey is said to be *analytical* (Oppenheim, 1992). In reality of course, even analytical surveys contain some descriptive variables, which are necessary to define the sample and to provide the independent or predictor variables.

Each question in the survey generates a variable, which is coded with (usually) a number assigned to each possible response. Although numbers are usually used for coding, it is vitally important to be aware of the type (or level of measurement) of each variable. Categories, such as 'procurement route', generate *nominal* variables, and if 'profitability' in our simple example is measured in actual sums of money, then this will be an *interval* or continuous variable. Interval variables that have meaningful zero points are known as *ratio* variables. The other types of variables are *ordinal*, which are created by ordered categories, such as those generated by attitude scales. The reason that we need to be aware of levels of measurement is so that we apply appropriate statistical analysis techniques to the data generated by our questionnaire.

In any questionnaire study, the data collection and data analysis stages are relatively straightforward; what takes the time is the design of the study, and it is this stage that the first part of this chapter will concentrate on. There have been very many studies in the social sciences that have failed because insufficient effort has been expended on their design. Good design is not merely a matter of ensuring that appropriate questions are posed to enable the aims of the study to be achieved, but also to ensure that an adequate response rate is achieved. Questionnaire fatigue is something that any researcher anticipating the use of postal questionnaires should be aware of, and achieving the best possible response rate is important to securing an adequate sample.

Questionnaires are useful research instruments in situations where 'snap-shots' or a limited amount of data are required and the same questions are to be asked of each respondent. An effective questionnaire can provide both reliable and valid information at a reasonable cost (Peterson, 2000). In situations where deeper and richer data are sought, and where more flexibility in the questions to be asked of different respondents is desired, then they are not suitable and the researcher will need to resort to other data collection methods.

Construction

As with the development of any research instrument it is imperative to commence its design with reference to the aims of the study. These may be expressed in terms of hypotheses or research questions but quite obviously these need to guide the questions being asked. Design commences with drawing up a list of questions that will enable the hypotheses to be tested or research questions to be explored.

Question types

Questions can seek factual information, ask about actual or likely behaviour, and explore the knowledge, attitudes and attributes of respondents. In terms of format, there are two types – open-ended and closed-ended. The latter have a fixed number of forced responses while open-ended questions can have an infinite number of responses. Open-ended questions are useful in exploratory research where respondents may suggest something that the researcher has overlooked. For this reason it is also useful to include such questions at the pilot stage, particularly if experts are used to pre-test the questionnaire. Closed-ended questions are however much easier to code and analyse.

Wording questions

Peterson (2000) suggests that questions should be worded with the following in mind:

- *Common sense* – as far as possible words should be phrased as one would in normal conversation.
- *Knowledge* – the researcher must have a passing knowledge of linguistics, cognitive psychology and communications science. In other words, questions must be phrased so that respondents understand them.
- *Experience* – the researcher must serve an 'apprenticeship' in question design before attempting an important study.
- *Brevity* – questions (and indeed the entire questionnaire) must be as brief as possible. Long questions that are difficult to understand will result in a poor response rate.
- *Relevance* – it goes without saying that questions should be relevant to the research aims but also those being surveyed must see the relevance of answering questions. If the sample see that the study is of relevance to them, they are more likely to respond.
- *Unambiguous* – Peterson recounts an interesting story of a survey being pre-tested on a colleague who was asked about the number of 'sewers' in the household. Readers may well reach the same conclusion as his colleague – that he was being asked about the drainage of the house. In actual fact the survey was enquiring about those living in the house who possessed proficiency with a needle and

cotton! Clearly if those being surveyed do not understand the questions posed, they are unlikely to provide sensible answers.

- *Specific* – the designer must be certain that the proposed sample has the knowledge to answer the questions posed.
- *Objective* – the important thing here is to avoid bias. It is imperative that leading questions are not posed.

The key concepts of reliability and validity also need to be considered at the design stage. Reliability is concerned with whether the instrument would produce the same results if the study was repeated with a similar sample, while validity is concerned with whether the survey is measuring what the researcher intended it to measure. These issues will be revisited in the second half of this chapter.

Structure

As implied above, an over-long questionnaire will not be completed and returned. If complex questions need to be posed then the questionnaire is not the appropriate research instrument to use. However, it is obvious that some questions will be more difficult to answer than others. The more difficult questions should be left until the end of the questionnaire in the hope that the respondent has by then invested some time and effort in completing the survey and will then not abandon it.

Attitude scales

Questionnaires are useful for recording the attitudes of respondents to particular statements or scenarios. They usually do this by employing a particular type of closed question – the attitude scale. There are many types of attitude scale but the two most commonly used are the semantic differential and the Likert scales.

The semantic differential scale

This is a widely used scale and was developed by Osgood *et al.* (1957). It usually has seven categories which are anchored between bi-polar labels. Thus a student of the author who was investigating the general public's attitudes to timber-frame housing, as opposed to traditional masonry construction, used a scale, part of which looked like Figure 11.1.

Notice that the third item of the scale has reversed poles and it is normal to use these with semantic differential scales to discourage respondents from merely ticking the same category for each item.

The Likert scale

The other most used scale was developed by Rensis Likert (Likert, 1932) and employs declarative statements and a list of response categories, typically five or seven.

Strong	____	____	____	____	____	____	____	Weak
Attractive	____	____	____	____	____	____	____	Ugly
Cool	____	____	____	____	____	____	____	Warm

Figure 11.1 Part of 'timber-frame housing' semantic differential scale.

A five-point scale for example may have 'strongly agree', 'agree', 'neutral', 'disagree' and 'strongly disagree' as the responses. Note however, that there is no reason why the responses should be balanced (Tull and Hawkins, 1984). The author has used an unbalanced scale when it was known that respondents had overwhelmingly positive views about particular statements. In this case the five responses ranged from 'very strongly agree' to 'disagree' (Hoxley, 1994, 1995).

Ordinal or interval data?

Strict interpretation of the rules of measurement requires that the data generated by such scales be treated as ordinal. However, some researchers (e.g., Weisberg et al., 1996) argue that provided the intervals between the various possible responses are approximately equal, then such data can be regarded as interval data, which of course enables more sophisticated statistical techniques to be undertaken. This is an important issue that needs to be explicitly addressed by any researcher who treats attitude scale data as interval data.

Multiple-item scales

Sometimes researchers wish to construct scales that have multiple items to model a particular construct. This is where several individual attitude scales are combined into a scale to measure some complex psychological construct. The discipline of psychometrics is devoted to multiple-item scale development and should be consulted by those wishing to undertake such studies. However, Peterson (2000) provides a concise summary of the development of such scales. Additionally, the section on factor analysis later in this chapter describes the construction and evaluation of a scale to measure clients' perceptions of the service quality provided by built environment professionals.

Piloting

Piloting or pre-testing is absolutely vital to the success of a questionnaire study. Research students will always pilot the questionnaire initially with their supervisor who will wish to ensure that there are no silly spelling or grammatical errors that could discredit the student or indeed the institution at which they are studying. Piloting proper should be with experts, in other words with the same type of person as will make up the main study sample. There may be a two-stage piloting process – firstly piloting the questions and then the questionnaire, but it is more likely to be carried out in a single stage for most small to medium-sized studies.

There needs to be some method by which the sample used for piloting can give feedback on their completion of the survey. This may be verbally when the piloting is carried out face to face. Where carried out at a distance, it may be that there needs to be a short questionnaire used to gain this feedback. Weisberg et al. (1996) suggest that for important studies the completion of the survey at the pilot stage could be audio or video-taped so that any difficulties with completion of the questionnaire are recorded and studied later, to help inform any necessary changes.

Sampling

In most surveys it would be impractical (not to say expensive) to collect data from the entire population being studied. It is thus necessary to select a sample to represent that population. The obvious questions the researcher needs to ask are: how will the sample be selected, and what should the sample size be? It is necessary to set up a sampling frame, that is, a list of cases that represents the population. Ideally, this would be a complete list of the population, but again, in many instances this would be impossible. In a study of UK Building Surveying firms, the author's sampling frame was the 500 or so firms that were listed in the RICS Directory (Hoxley, 1994, 1995). Selecting the sample from the sampling frame can be carried out using probability or non-probability methods. Probability sampling assumes that every part of the sample has an equal calculable probability of being selected. Thus in the example given above, the probability was 0.5 since every other firm in the sampling frame list was selected for the sample. In non-probability sampling, the researcher employs other criteria for selection. These may be firms that he or she knows, or firms that are thought to be experts on the subject under consideration.

In terms of the size of the sample, it is necessary to consider the sampling accuracy that the researcher considers acceptable. Of course, it is also necessary to think of any analysis of subgroups and whether the size of the sample will be acceptable for the particular analysis to be performed on each subgroup. Sample size is a very complicated issue but Weisberg et al. (1996) provide a good summary of the issues to be considered.

Administration

Postal surveys

The traditional way to gather survey data for research has been by postal questionnaire. When administering postal questionnaires great care has to be taken to ensure an adequate response rate. Typical response rates quoted in text books have a mean of about 30 per cent but one has to work hard to achieve this level of response. When the author first used a postal questionnaire (in 1992) he achieved very nearly a 70 per cent return rate (Hoxley, 1994, 1995). The case study presented at the end of this chapter achieved a response rate of nearly 50 per cent but when a follow-up study, using almost identical instruments sent to a similar sample, was conducted some eight years later, a response rate of only 19 per cent was achieved (Hoxley, 2007). These examples highlight just how difficult it is to achieve a sample of adequate size using postal administration. Steps that can be taken to achieve the best possible completion rate include sending the questionnaire out with an individually addressed letter explaining the purpose and relevance of the study. This involves compiling a database of these individuals and using a 'mail merge' facility to create the letters. A stamped addressed envelope should be sent out with each questionnaire – even though a majority of these will end up in the waste paper bin! It may be necessary to send follow-up letters to non-respondents. This is also evidence (Berdie et al., 1986) that using a term other than 'questionnaire' also increases the response rate. One of the author's students sent out a postal questionnaire with a wrapped sweet and told the recipients that by the time they had finished eating the sweet, they would have completed the questionnaire. Another way of encouraging a response is to promise feedback on the results of the research. It goes without saying, however, that if promised, such feedback must be provided. Nowadays, it is relatively easy to provide this, simply by asking those interested in receiving feedback to give their email address on the questionnaire. It is then a simple task to email the final research report, or a summary of it, to those requesting feedback. Clearly the anticipated response rate needs to be factored into the decision over the number of questionnaires to be sent out in order to achieve the minimum sample size required for the study.

Verbal methods

When the general public is being surveyed, an easier way to administer the questionnaire is face to face. Such surveys can be conducted in the street provided that adequate security considerations are taken. Surveys can also be conducted by telephone but there is obviously the cost of the telephone calls to be considered. Such surveys need to be fairly short since respondents to such surveys tend to be impatient if the survey lasts for more than a few minutes. Cold calling is not a good idea and some contact should be made in advance to arrange a convenient time to conduct the telephone survey.

Increasingly, of course, the internet is being used to administer surveys. While these can be conducted by email, a rather better method is to have the survey set up on a web-page with a link sent to respondents by email. Various software packages, some commercially available but others produced by academic institutions for the use of their students, are available to assist in the production of web-based survey forms.

Coding

Coding is the allocation of (usually) a number to each possible response to each question. Clearly this is rather easier for closed-end questions than for open-end questions. Although possible responses to open-end questions can be predicted, it is probably best to wait until the data entry stage before coding such responses. It is usually necessary to group similar answers with the same code, since no two respondents are likely to give exactly the same answer to such questions. A closed-end question such as a five-point Likert-scale has six possible codes – one for each response and another for failure to provide a response (i.e. a missing value). It is most important that the analysis framework is set up before the data are collected. Indeed many researchers would advocate entering dummy data and running the complete statistical analysis before the survey. Such actions will help to iron out any problems before the real data are collected.

Software packages

SPSS is the most used statistical analysis software and is extremely powerful – as well as carrying out the full range of statistical procedures, its chart drawing facility is excellent. It is a Windows-based environment and it is fairly straightforward to set up a data entry spreadsheet and to analyse data and draw charts. There are tutorials to help the researcher determine which is the most appropriate technique to use, but it is necessary for the researcher to have some basic knowledge of statistics and to be prepared to learn more about the particular techniques required for their investigation.

Coding missing values

The treatment of missing values needs to be given careful consideration. For most analysis undertaken in SPSS, the default position is that any cases that have any missing value for any of the variables under consideration will be omitted from the analysis. This can lead to results other than those desired by the researcher. Thus, if a respondent fails to provide an answer to one item of a multiple-item scale then any subsequent analysis (e.g., to assess its reliability or to carry out factor analysis) will automatically omit all of the scale variables for that particular case if the SPSS default for missing values is followed. In such situations it is sensible to select the option that replaces any missing value with the mean value of that variable.

Data entry

Data entry can be carried out by keying into a spreadsheet manually or, if an optical readable questionnaire is used, data input can be carried out automatically. The author and a colleague collected data from a large sample of surveyors about the equipment they carry with them when carrying out condition surveys of buildings (Hoxley and Coday, 2002). An optically read questionnaire was used and input of all data into SPSS took less than an hour, which is a fraction of the time it would have taken to manually key in the data. As well as time savings, another advantage of using an optical reader is that it completely removes any possibility of making errors that can arise when manually entering data. Of course, the researcher must have access to expertise to create the data collection form and to the hardware to read the forms. However, most universities use this technology to read their module evaluation student questionnaires, so it is usually available.

In the following section the useful statistical technique of factor analysis is introduced by considering a built environment case study.

Factor analysis

Factor analysis is a statistical technique for aggregating many variables into a few underlying factors, dimensions or constructs. Say, for example, that we are interested in the reasons used by undergraduate applicants to select the universities to which they apply. We would first generate a list of say 10–20 variables that we think may be important to their choice. These may include teaching quality rating, research standing, employability, reputation for an active night-life, sports facilities, geographical location and cost of living. We could then get several hundred first year undergraduates to complete our multi-item scale to discover what had informed their recent choice. An exploratory factor analysis could then be performed to condense these 10–20 variables into a few more meaningful factors. It might be that the results of this analysis would reveal that there were three factors that we could identify as 'academic', 'social' and 'economic'. This example is said to be 'exploratory' since we are investigating something we are unsure about. Where factor analysis is used to compare a result with one that has been hypothesised, then it is said to be 'confirmatory'. Since most factor analysis tends to be exploratory we will concentrate on this type.

To demonstrate the technique we will consider the author's development of a scale that was designed to measure service quality in a built environment professional context. The original scale had 28 items which largely originated from relevant literature. The principal source was the highly influential study of a group of US researchers (Parasuraman et al., 1988, 1991), the main output of which was a generic service quality measurement scale known as SERVQUAL. Other literature consulted were studies of real estate, architectural and building surveying service quality. In selecting the items of any multiple-item scale the researcher should ensure that there are sound theoretical underpinnings to their inclusion (Churchill,1979).

A questionnaire was then designed for clients to rate, using a seven-point Likert scale for each of the 28 items, their perceptions of service quality received from a professional consultant recently. The consultant was not identified but was merely referred to as 'XYZ'. The statements of the scale and their corresponding variable names may be seen in Table 11.1.

Data collection

The scale was pre-tested by visiting and interviewing senior personnel employed by the property departments of six organisations. The organisations (an American fast

Variable name	Statement
TECH	XYZ use up-to-date technology
OFFICES	The offices of XYZ are visually appealing
STAFF	The staff of XYZ are always tidy in appearance
PRESENT	The written and graphical output of XYZ is well presented
SIZE	XYZ's size is appropriate for the services they perform for me
CORRECT	XYZ's solutions to problems are technically correct
DESIGN	The design element of XYZ's work shows creativity and capability
TIME	XYZ provides its services at the time it promises to
WHEN	XYZ tells me when it will perform the service for me
PROMPT	XYZ provides prompt service
WILLING	XYZ and its employees are always willing to help me
BUSY	XYZ and its employees are never too busy to respond to my requests
ACCESSBL	Employees of XYZ are easily accessible to me
SAFE	I feel safe in my dealings with XYZ
POLITE	XYZ and its employees are always polite to me
COMPETEN	Employees of XYZ have the knowledge and competence to solve my problems
EXPERIEN	XYZ and its employees have experience relevant to the service I require
PERSONAL	XYZ provide me with personal attention
BESTINTS	XYZ have only my best interests at heart
UNDERSTA	XYZ understand my problems
LONGTERM	I will benefit from a long-term working relationship with XYZ
SIMILAR	XYZ and I have similar views about things that are important
COSTCONT	XYZ provide good cost control of projects
INVOLVED	The partners or directors of XYZ stay involved with my projects
SITESUPV	The site supervision of projects by XYZ is good
LOCATION	XYZ's offices are conveniently located for me
VERBALPR	The standard of verbal presentation by employees of XYZ is good
UNDERORG	XYZ and its employees understand my organisation

Table 11.1 Scale statements and variable names.

food chain, a national electricity generator, a hospital NHS trust, a metropolitan borough council, a university and a central government department) all completed the questionnaire, indicated that they would have completed it had they received it through the post and they also provided further useful feedback that resulted in minor amendments being made. A questionnaire was posted, together with a covering letter and a stamped addressed envelope, to a named senior employee in 500 client organisations located throughout the UK. Two hundred and forty-four questionnaires in a useable form were returned, representing a 48.8 per cent response rate.

Correlation matrix

The factor analysis commenced with a study of the correlation matrix of all 28 of the original scale variables. Hedderson (1991) suggests that any variable whose correlations with the other variables are less than 0.4 in absolute terms should be excluded from the factor analysis. This is because variables in a scale are generally attempting to measure some aspect of the same underlying construct and any variables that are not correlated with others should be ignored. There were two variables that fell into this category and both were concerned with the professional firm's office premises (with its *appearance* and *location*). The correlation matrix suggests that neither of these variables is important to the clients of the professionals assessed and these two variables were excluded from the scale at this stage of the analysis.

Sampling accuracy

It is then necessary to examine the data to see whether it is suitable for factor analysis. This involves calculating various measures of sampling accuracy. Bartlett's test of sphericity (which tests the hypothesis that the matrix is an identity matrix – that is all diagonal terms are 1 and all off-diagonal terms are 0) was 2284 with an associated very low significance level. This suggests that the correlation matrix is unlikely to be an identity, in other words relationships are likely to exist between variables. Another indicator of the strength of the relationship among variables is the partial correlation coefficient. If variables share common factors, the partial correlation coefficients between pairs of variables should be small when the linear effects of the other variables are eliminated. The Kaiser–Meyer–Olkin measure is an index for comparing the magnitudes of the observed correlation coefficients to the magnitudes of the partial correlation coefficients and for the original correlation matrix was 0.93. The negative of the partial correlation coefficient is called the anti-image correlation. The anti-image correlation matrix was computed and the smallest measure of sampling accuracy was 0.90. All of these results suggest that the data collected were suitable for factor analysis (Norusis, 1994).

Extracting factors

The remaining 26 variables of all 244 cases were then subjected to principal-components analysis, which is a procedure which *extracts* the factors. The first principal component is the combination that accounts for the largest amount of variance in the sample. The second component (uncorrelated with the first) explains the next largest amount of variance, and so on. Principal-components analysis is the main method of extraction used in factor analysis but there are others. (For a fuller discussion see Bryman and Cramer (2001). This text is very useful as it takes the reader through the SPSS menus for carrying out a factor analysis.) In this case the principal components analysis extracted four

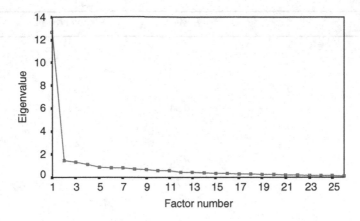

Figure 11.2 Factor analysis scree plot.

factors which together accounted for 64 per cent of the variance. The amount of variance that each factor accounts for is known as its eigenvalue and the convention is to retain only those factors that have an eigenvalue of greater than one. The remaining 22 factors only explained the remaining 36 per cent of the variance and this suggests that a four-factor model fits the data collected. Since Factor 1 explains 48.8 per cent of the variance it could be argued that service quality is a uni-dimensional construct and reference to Figure 11.2 which is a scree plot of the factor analysis eigenvalues supports this view.

Rotation

The next stage in factor analysis is to *rotate* the factor matrix. This is a procedure which attempts to identify the factors. After rotation the number of larger and smaller factor loadings increases, that is variables are more highly correlated with single factors and more meaningful interpretation of the factors becomes possible. The method of rotation selected was *oblique* which allows for correlations between factors, as opposed to orthogonal rotation which assumes no correlation between factors. It is unlikely that the factors are completely uncorrelated and 'oblique rotations have often been found to yield substantively meaningful factors' (Norusis, 1994). Oblique rotation was used in the development of the original SERVQUAL scale (Parasuraman *et al.*, 1988). Rotation is an iterative process and the data converged in 20 iterations.

Identifying factors

Naming the factors is a question of looking at the pattern matrix following rotation and looking for connections and similarities for those variables which load on particular factors. Interpretation of the pattern matrix resulting from the rotation phase of this analysis, which is reproduced in a simplified form in Table 11.2, suggests that the four factors could be titled 'What', 'When' 'How' and 'Who'. As will be seen, seven of the twenty-six variables loaded on more than one factor and Factor 1 is the most heavily loaded factor. Since this factor appears to be concerned with 'what' the professional actually provides in his or her service to the client, this is not surprising. The 'how' dimension of the analysis research (Factor 3) is concerned with the written and verbal presentation of the professional and such things as the technology employed and the appearance of staff. The only variables which load heavily on the 'when' factor and

Variable[b]	F1: 'WHAT'	F2: 'WHEN'	F3: 'HOW'	F4: 'WHO'
TECH			0.79	
STAFF			0.50	
PRESENT	0.27		0.64	
SIZE		0.54	0.33	
CORRECT	0.81			
DESIGN	0.38		0.44	
TIME		0.75		
WHEN		0.78		
PROMPT		0.77		
WILLING				0.68
BUSY				0.71
ACCESSBL				0.79
SAFE	0.45			0.45
POLITE				0.80
COMPETEN	0.82			
EXPERIEN	0.67			
PERSONAL			0.26	0.55
BESTINTS	0.55			
UNDERSTA	0.70			
LONGTERM	0.67			
SIMILAR	0.83			
COSTCONT	0.67	0.26		
INVOLVED	0.37			
SITESUPV	0.59			
VERBALPR			0.47	0.29
UNDERORG	0.53			

[a] Loadings of 0.25 or less are not shown.
[b] See Table 11.1 for full description of the variable.

Table 11.2 Factor loadings[a] of variables.

which are not directly related to time, are the size of the professional's firm and cost control of projects. However, the firm's ability to deliver a service on time is obviously not unrelated to its size and time is an important aspect of cost control. The 'who' factor is mainly concerned with the 'people issues' of the service provision, for example their willingness to help, how busy they are, how accessible they are and whether they are polite to clients.

Scale reliability

A measurement scale such as the one developed during this research must be both *reliable* and *valid* (Churchill, 1979). Reliability is concerned with the internal consistency of the scale, that is: does the scale behave similarly when administered by

Dimension	Factor	Number of items	Cronbach's Alpha
WHAT	F1	13	0.95
WHEN	F2	5	0.86
HOW	F3	7	0.81
WHO	F4	7	0.90
ENTIRE SCALE	ALL	26	0.96

Table 11.3   Internal consistencies of the scale.

different people? The most widely used reliability coefficient is Cronbach's Alpha (Cronbach, 1951), which can range from 0 to 1 with higher figures indicating greater reliability. The results of the computation of Alpha for each factor and for the scale as a whole are presented in Table 11.3.

These figures are all high and at a generally higher level than for the original SERVQUAL scale (Parasuraman *et al.*, 1988). The total scale alpha figure of 0.96 suggests that the scale has very good reliability.

Scale validity

In order to test whether the scale does what it is intended to do (that is, measure the service quality of the professionals assessed) a score was computed for each case. This score consisted of the mean score of all variables for which the client gave an assessment of the professional. The lowest possible score is 1.0 (for a professional rated as providing a very much worse service than expected for every variable) and at the other end of the assessment spectrum the highest possible score is 7.0. In fact, the lowest score recorded was 2.74 and the highest was 6.15. Separate one-way analyses of variance were then performed between these computed scores and responses to questions about the overall quality rating and whether the client would recommend the professional to another organisation. This procedure aims to establish whether the scale score is capable of distinguishing between the responses to these other questions. For both questions, the scale score was successful in distinguishing between groups and both analyses resulted in high F ratios with very small associated probabilities. These analyses of variance results suggest that the scale possesses construct validity.

Thus, the developed scale is both a reliable and valid instrument to measure service quality in a construction profession context. For a fuller description of the development of this multiple-item scale see Hoxley (2000a). The factor analysis technique allows researchers to gain insights into complex psychological issues and, in marketing situations, to target resources at the most important factors revealed by the analysis. For example, in the university example cited earlier, if it transpired that the 'economic' factor was the most heavily loaded factor then the university could target the variables that are loaded on that particular factor. The multiple-item scale developed by the author was used to investigate various hypotheses concerning fee tendering and service quality (Hoxley, 2000b).

Summary

This chapter has considered the design, administration, data entry and data analysis procedures employed when using questionnaires as research instruments. It is necessary to be aware of the type of data collected and to use appropriate statistical techniques depending upon whether interval, nominal or ordinal data have been collected. Although software packages such as SPSS have made analysis much easier than in the past, it is necessary for the researcher to have some awareness of statistics to ensure that the appropriate technique is used. A case study of the use of factor analysis in a built environment context has also been presented.

References

Berdie, D.R., Anderson, J.F. and Niebuhr, M.A. (1986) *Questionnaires: Design and Use*, Scarecrow Press, London.

Bryman, A. and Cramer, D. (2001) *Quantitative Data Analysis*, Routledge, Hove.

Churchill, G. (1979) A paradigm for developing better measures of marketing constructs, *Journal of Marketing Research*, 16, 64–73.

Cronbach, L.J. (1951) Coefficient alpha and the internal structure of tests, *Psychometrika*, 16, 297–334.

Hedderson, J. (1991) *SPSS/PC+ Made Simple*, Wadsworth Publishing Company, Belmont, CA, USA.

Hoxley, M. (1994) Assessment of building surveying service quality: Process or outcome?, *RICS Research Series Paper*, 1(8).

Hoxley, M. (1995) How do clients select a surveyor?, *Structural Survey*, 13(2), 6–12.

Hoxley, M. (2000a) Measuring UK construction professional service quality: The what, how, when and who!, *International Journal of Quality and Reliability Management*, 17(4/5), 511–526.

Hoxley, M. (2000b) Are fee tendering and construction professional service quality mutually exclusive?, *Construction Management and Economics*, 18(5), 599–605.

Hoxley, M. (2007) The fee tendering and service quality issue revisited, *Property Management*, 25(2), 180–192.

Hoxley, M. and Coday, A. (2002) Portable test equipment for building surveys – a study of current practice, *Proceedings of the RICS Construction and Building Research Conference*, Nottingham Trent University, September, pp. 197–208.

Likert, R. (1932) A technique for the measurement of attitudes, *Archives of Psychology*, 140, 5–55.

Norusis, M.J. (1994) *SPSS Professional Statistics 6.1*, SPSS Inc, Chicago, USA .

Oppenheim, A.N. (1992) *Questionnaire Design, Interviewing and Attitude Measurement*, Continuum, London.

Osgood, C.E., Suci, G.J. and Tannenbaum, P. (1957) *The Measurement of Meaning*, University of Illinois Press, Urbana, IL.

Parasuraman, A., Zeithaml, V.A. and Berry, L.L. (1988) SERVQUAL: A multiple-item scale for measuring consumer perceptions of service quality, *Journal of Retailing*, 64(1), 12–40.

Parasuraman, A., Zeithaml, V.A. and Berry, L.L. (1991) Refinement and reassessment of the SERVQUAL scale, *Journal of Retailing*, 67(4), 420–450.

Peterson, R.A. (2000) *Constructing Effective Questionnaires*, Sage Publications, London.

Tull, D.S. and Hawkins, D.I. (1984) *Marketing Research Measurement and Method*, Collier MacMillan Publishers, London.

Weisberg, H.F., Krosnick, J.A. and Bowen, B.D. (1996) *An Introduction to Survey Research, Polling and Data Analysis*, Sage Publications, London.

Chapter Twelve
Using software to analyse qualitative data

Andrew King

Introduction

The field of qualitative research is undergoing a mini-technological revolution fuelled heavily by the proliferation of Computer Assisted Qualitative Data Analysis Software (CAQDAS) (Richards, 2005). Various packages are available and are increasingly being used owing to their ability to store, organise and manage the fracture and analytic reassembly of large amounts of data, whilst offering high levels of transparency. Their wealth of features, stability, speed and ultimate power lie in stark contrast to similar yet unsophisticated traditional manual techniques such as highlighted hard copy coding, copied extract theme building and manual frequency counts. However, CAQDAS should not be viewed as a panacea which automatically produces research that is rigorous or systematic (Lee and Fielding, 1996; Weitzman, 2000; Blismas and Dainty, 2003).

The adoption of this type of technology has been accompanied by various difficulties and this chapter focuses on three such major problems. First, the relationship between software and methodology is discussed. Not only is CAQDAS confused as an overall methodology, but also CAQDAS is still viewed by many as separate to the analysis process (Fielding, 2002; Weitzman, 2003). This understanding lies in stark contrast to the practicalities of using CAQDAS in the research process; the two are symbiotically related. Second, issues surrounding analytic distance; one of the most common outcomes of uninformed use of the software, allied to an understandable wish to seek greater closeness to the data, is the tendency to 'over-code' in an unconscious manner often leading to confusion and a lack of credibility and trustworthiness in the final research product (Di Gregorio, 2003). Third, there is a paucity of software training and support which provides sufficient depth of study and recognises the need to integrate CAQDAS and overall qualitative methods training (Carvajal, 2002; Johnston, 2006). This lack of understanding leads to short-term training courses focused around the technological aspects of the software (Carvajal, 2002). These three issues impact on CAQDAS's ability to improve the quality of qualitative research.

This chapter draws on the author's experiences of using various CAQDAS software packages on a number of projects. Instead of proposing a didactic recipe book of advice to those developing their own approach to using CAQDAS, this chapter provides an introduction to the topic by drawing on the key authors in the field, synthesising their works and reflecting on the challenges and potential pitfalls faced by researchers. In so doing, it stresses the need to integrate technical and methodological skills and avoid elevating CAQDAS beyond its status as a tool to be used in the overall research process.

Why use software?

> … the researcher who does not use software beyond a word processor will be hampered in comparison to those who do.
>
> (Miles and Huberman, 1994, pp. 43–44)

Few today would disagree with the above statement, indeed the use of CAQDAS has increased significantly over recent years and its rise in popularity is perhaps not surprising when one considers the benefits associated with the wide ranging choice of current software. Traditionally, qualitative researchers have manually conducted the fracture and analytic reassembly of qualitative data using various processes; readers are advised to refer to Tesch (1990), Miles and Huberman (1994), Denzin and Lincoln (2000) and Welsh (2002) for further reading. Some examples include:

- Manual frequency counts of words or collections of words.
- Hard copy coding, which generally includes transcribing interviews, copying extracts and coding via highlighting sections followed by copied extract theme building which entails reforming extracts in themes via pasting thematically congruent pieces of paper together.

In general, such manual processes are time consuming, messy, unstable, complicated and often lethargy inducing. CAQDAS, on the other hand, automates many of these processes and many more as shown in Table 12.1.

The way CAQDAS increases the speed with which tasks can be carried out is often quoted as one of its key advantages (Carvajal, 2002; Blismas and Dainty, 2003; Richards, 2005). However, for Bourdon (2002), increased speed brings other benefits; faster means differently. He points out that the way in which CAQDAS allows previously prohibitive tasks to be carried out almost instantly provides researchers with more analytic options. For example, the ability to allow hunches to be tested without using significant amounts of time and resources is cited as a major advantage. Whilst the raft of benefits made available by software is easy to recognise, users need to be conscious of various problematic issues surrounding usage, which will be discussed later in the chapter. First, we consider the range of software available.

Comparison of software

Early debate in the use of qualitative software was dominated by comparisons of software packages (Kelle, 1995; Fisher, 1997). Today, there is a considerable range of software each with its own distinctive features (Lewins and Silver, 2007). No single software package can be considered to be the 'best' for a range of methodologies (Koenig, 2004) and new users will be best served by conducting an exploration of each offering in light of their particular methodological needs. It would be all too easy, and most ill advised, to select a package owing to its perceived or real popularity or similarly simply basing one's decision-making solely on previous experience of a particular package.

A review of software falls outside the scope of this chapter. However, the following list, whilst not claimed to be exhaustive, provides readers with details of the main packages and contact details for further exploration. Readers are advised to refer to Lewins and Silver (2007) and the CAQDAS Networking Project (http://www.soc.surrey.ac.uk) for further information.

- ATLAS.ti Version 5. http://www.atlasti.com
- HyperRESEARCH Version 2.6. http://www.researchware.com
- InfoRapid. http://www.inforapid/html/english.htm

- Relationships – Specifying relationships among codes and using these relationships in the analysis.
- Memos – Writing memos and enabling these to be linked to text and codes.
- Hypertexting – Creating links between different points in the text.
- Multimedia – Enabling the use of audio and video media.
- Mixed methods – Linking qualitative and quantitative data including statistical packages.
- Notes – Making notes in the field and writing up/transcribing notes.
- Editing – Correcting/extending/reviving field notes.
- Coding – Attaching key words/tags to segments of text/graphics/video/audio to allow for later retrieval.
- Storage – Keep text in an organised manner.
- Search and retrieve – Locate relevant segments of text and enable them to easily be inspected.
- Data linking – Connecting relevant data segments to each other, format categories, clusters or networks of information.
- Memoing – Write reflective commentaries on some aspect of the data, theory or method as a basis for deeper analysis.
- Content analysis – Count frequencies, sequences or locations of words or phrases.
- Data display – Place selected or reduced data in a condensed, organised format such as a matrix or network for inspection.
- Conclusion drawing and verification – Aid in the interpretation of displayed data and the testing or confirmation of the findings.
- Theory building – Developing a systematic, conceptually coherent explanation of the findings/testing hypotheses.
- Graphic mapping – Creating diagrams that depict findings or theories.
- Report writing – Interim and Final.
- Research journal – Develop a journal that can be fully linked to other documents, nodes, memos and store in one secure place.
- Literature review – Import literature and work in a live environment enabling more seamless links between data, ideas and literature.

Table 12.1   CAQDAS features (adapted from Weitzman, 2000).

- Kwalitan Version 5.0. http://www.kwaitan.net/engels
- MAXQDA 2007. http://www.maxqda.com
- QDA Miner Version 3.0 and WordStat 5.1. http://www.provalisresearch.com
- QSR Nvivo Version 7. http://www.qsrinternational.com
- QUALRUS Version 4.0. http://www.qualrus.com
- Storyspace 2. http://www.eastgate.com
- TRANSANA Version 2.20. http://www.transana.org

Methodology and software

Computers make good friends. No matter how stupid, dull or dumb we may feel, we can still feel smarter than our computer. Computers can do many things, but they cannot think – and we can. Unfortunately, that also means the thinking is up to us. A computer can help us to analyse our data, but it cannot analyse our data. This is not a pedantic distinction: we must do the analysis.

(Dey, 1993, p. 55)

Whilst CAQDAS undoubtedly provides us with a quick, powerful and secure way to store, search and help analyse data, the research community must remain ever mindful that software does not by itself either provide an overarching methodology or carry out the analysis of qualitative data by itself. The golden rule that should always be borne in mind is that it is the *researcher who does the analysis; not the software itself.*

Concern has been raised that CAQDAS can carry out the analysis itself (Fielding and Lee, 1991), interfere with wider methodological considerations (Agar, 1991; Seidel and Kelle, 1995), potentially guide researchers in a particular research direction and could, owing to its close relationship with grounded theory (Lonkila, 1995), lead to a new orthodoxy of qualitative analysis (Coffey *et al.*, 1996). Whilst the criticisms of grounded theory's unhealthy relationship with CAQDAS have been largely dismissed by Lee and Fielding (1996) and Kelle (1997), the relationship between the way many types of CAQDAS focus on data fracture and reassembly via the code and retrieve cycle is largely in tune with grounded theory's (Glaser and Strauss, 1967; Strauss and Corbin, 1998) coding process and readers are advised to refer to the work of Bringer *et al.* (2006) for an eloquent and well-rounded illustration of just such an approach. Nevertheless, such similarity should not blind users into simple correlation between CAQDAS and a single methodology; CAQDAS can be used for a range of methodologies from discourse analysis, through ethnography, hermeneutics and beyond. Underlining this point, Lee and Fielding (1996) took a sample of qualitative research studies employing CAQDAS and found that around 70 per cent of studies had no direct relationship to grounded theory. Kelle (1997) believes the relationship between CAQDAS and grounded theory may be a product of two issues. First, software developers need to utilise a methodological framework, and grounded theory is a readily recognisable brand name. Second, grounded theory has a relatively unique status in qualitative research as those using the methodology tend to describe the analysis process in great detail.

The important point to consider is the propensity for the software to interfere with the overall methodological design, which should always be led by the research aims and resultant questions. In terms of the overriding research design, it is the ontological, epistemological, methodological and method(s) that should inform the design; not the myopic selection of CAQDAS. Bringer *et al.* (2004), contemplating the temptation for researchers to substitute CAQDAS for an overriding methodological framework in order to demonstrate the rigour of their work, stresses the need for researchers to prove how CAQDAS fits within the overall research design.

Blismas and Dainty (2003) recognised the increased popularity of CAQDAS in the construction management (CM) research community and the accompanying lack of debate surrounding its use. Similar to other authors, for example Bringer *et al.* (2004), they believe that CAQDAS can have a negative impact on research where users believe it to be a panacea which ensures rigour and transparency. Their calls for increased debate in this area are laudable, as the general lack of CAQDAS-related debate in qualitative research circles in general (Richards, 2002, 2004) is magnified in the applied field of CM and the wider study of the built environment.

Pointing to the possibility that CM-based qualitative research could be reduced to a sterile convergence of CAQDAS-based methods and approaches, Blismas and Dainty (2003) state that 'many CM researchers now turn to CAQDAS as the standard method for analysing textual data without considering implications to their entire research design' (p. 462).

Taking account of the discussion of the real and important need to develop an approach to using CAQDAS that is grounded in a much wider understanding of qualitative research methodology, it is worth pausing to consider how CAQDAS can be integrated into the wider research project. Richards (2005) provides significant assistance in this respect and readers are strongly advised to refer to her work, principally *Handling Qualitative Data*.

Analytic distance

There has been significant early debate relating to the computer's propensity to stop qualitative researchers getting close to their data (Seidel, 1991; Barry, 1998; Fielding and Lee, 1998; Weitzman and Miles, 1995). The view that computers create analytic distance is associated with the belief that the automation of various processes that were previously carried out manually, and required one to 'work with the data', diminish one's ability to get to know the data. My own experience is that it is more effective to initially work with hard copies of interview transcripts to kick-start the process of getting close to the data. Interestingly, current concerns are centred on the problem of becoming too close to the data via the overuse of coding without the much needed distance to reflect and consider the bigger picture (Gilbert, 2002; Richards, 2005; Johnston, 2006). Becoming too close to the data is something that is generally termed the code and retrieve cycle, coding trap or coding fetishism (Gilbert, 1999; Richards, 2005; Johnston, 2006). Johnston (2006) associates mechanical coding of large parts of data with a lack of use of some of the software's tools, particularly search tools, which can help to 'see the proverbial wood from the trees' (p. 383).

Gilbert (1999) offers a useful way to conceptualise the analytic distance associated with CAQDAS. He found three levels of closeness to the data by interviewing qualitative researchers experienced in using both manual and computer-based analysis. First, the *Tactile-Digital Divide* where researchers work on screen rather than on a hard copy. Second, the *Coding Trap*, as described above, where researchers get too close to the data and lose perspective and third, the *metacognitive shift* where researchers reflect on the processes. He found that confident use of the software led users to undergo the metacognitive shift, at which stage they could correct issues such as overcoding. Johnston (2006) and Gilbert (2002) emphasise the importance of the research journal in gaining analytic distance by using it to record key decisions, reflections and emerging ideas and this advice is advocated.

Learning to use CAQDAS

Training researchers to use software is an area that has become increasingly important over recent years as witnessed by the number of authors who have tackled it as a central theme in their work (Carvajal, 2002; Johnston, 2006). Carvajal (2002) undertook research based on a sample of 44 CAQDAS training workshops identified through the CAQDAS Networking Project QUAL-*Software* mailing list. He found that the majority of sessions were one day in duration, did not require attendees to be qualitative researchers and, perhaps surprisingly, attendees did not require any previous knowledge of qualitative research. The workshops were relatively 'hands on' yet at the same time provided little opportunity for trainees to use their own data. The stated aim was, expectedly, to train attendees in the basic tools of the programme, yet only a few of the workshops looked at the relationship between the wider field of qualitative research and CAQDAS. Taking account of the increasing importance of CAQDAS there is a real need to address these issues. The following lessons, learnt through sustained use of CAQDAS over the past few years, may help to develop a more effective approach to getting the most from CAQDAS training in addition to aiding inter-coder reliability (Miles and Huberman, 1994) and are particularly appropriate to PhD students:

(1) Where possible, including a member of the supervisory team or a co-worker in the overall CAQDAS learning experience helps to develop similar knowledge reference points that can be repeatedly returned to and developed throughout the course of the research.

(2) Wider dual-reading, with a member of the supervisory team or a co-worker, of methodological literature will help to develop a more rounded rigorous methodology which addresses ontology, epistemology, theoretical frameworks, methodology and specific methods. Such comprehensive reading, whilst time-consuming and requiring sustained effort, will help to develop an informed approach to CAQDAS which avoids elevating it above its role as a methodological tool and not as a methodology in itself.

(3) Providing a member of the supervisory team, or a co-worker, with shared access to the CAQDAS-based project provides tangible benefits throughout analysis. For example, it allows supervisors to interrogate the work in much greater detail leading to much more productive supervision sessions.

(4) The concurrent development of the thesis, or other research product(s), and analysis records within the CAQDAS project, including their incorporation in supervision sessions, helps to ensure reliable interlinked records are available to refer to when writing up the thesis.

The main point here is the need to ensure CAQDAS training is integrated into a wider methodological training, and not, as is currently the case, separated. Johnston (2006) believes current CAQDAS training has led to two separate learning curves: the *technical* curve and the *methodological* curve. In her view, this situation has led to problems as students do not get the support they require from methods literature, postgraduate training and their supervisory team. Reflecting on a wealth of experience in the area, Johnston poses an important question when she asks 'Is it acceptable for doctoral supervisors and examiners to know less about computer-assisted approaches to analysis and the current methods revolution than their students?' (p. 383).

This is a difficult question, however, at the very least supervisors should have an appreciation of the role CAQDAS plays in the overall research process. Increasing the opportunities for supervisors to be formally trained in CAQDAS in tandem with ongoing sustained use of the software must become a priority if we are to realise the increased rigour that can undoubtedly be made available by today's advanced software systems.

The quality of qualitative research

CAQDAS has generally increased the debate surrounding quality in qualitative research as highlighted by the three major points discussed in the previous sections. In addition to the discussions of the way CAQDAS has increased the popularity of qualitative research generally (Fielding and Lee, 2002), the debate has stemmed from the software's ability to increase the transparency of the analysis process.

Whereas manual analysis is associated with a focus on the research product itself, in the form of a thesis or other outputs, CAQDAS brings with it an increased ability to assess the entire research process. Johnston (2006) believes that this transparency has the ability to increase the expectations on PhD students and she cites recent research (Spencer *et al.*,

2004) focused on improving the quality of qualitative work by laying out guidelines for quality assessment. Four principles were formulated, requiring work to be:

(1) Contributory.
(2) Defensible in design.
(3) Rigorous in conduct.
(4) Credible in claim.

Nevertheless, even though the work stresses the important role transparency plays in quality, and provides a useful working definition, the work fails to incorporate any references to CAQDAS. Clearly, the potential CAQDAS offers to increase the quality of qualitative research needs to be harnessed by the research community. Open debate and robust evaluation of the relationship between software, methodology and the final product is needed to reap the rich rewards available.

Conclusion

This chapter has highlighted the way CAQDAS can manage qualitative data and in so doing stressed the benefits that are made available in comparison to manual techniques. Such improvements have been contrasted with the more problematic effects in the shape of confusions over methodology and software, analytic distance and the problems associated with learning to use CAQDAS. Various ways of reducing these problems have been proposed in an attempt to enable readers to develop an informed effective approach to using CAQDAS. It has been made clear that, above all, the purpose of the research should not be forgotten; something that can prove difficult when it is so quick and easy to develop hundreds of codes. CAQDAS is merely a tool in the qualitative researchers toolbox and how it is used should be considered in light of the wider research design which in turn is driven by ontological, epistemological and methodological assumptions.

CADAS offers qualitative research a very important modern tool because it allows researchers to deal with more data in an interactive secure systematic and efficient fashion. However, it is important to remember that CAQDAS is not an end in itself and if users become overly focused on the software itself in a misguided attempt to find a 'right' way of analysing data, it is likely that they will become entrenched in the detail and potentially reduce the ability to build theory. As building theory is the ultimate goal of most qualitative research we need to ensure that we continue to operate at a deeper level and consider the methodological and philosophical issues relating to the research design.

References

Agar, M. (1991) The right brain strikes back, in Fielding, N.G. and Lee, R.M. (eds) *Using Computers in Qualitative Research*, Sage, London, pp. 181–194.

Barry, C.A. (1998) Choosing qualitative data analysis software: Atlas/ti and Nudist compared, *Sociological Research Online*, 3(3). http://www.socresonline.org.uk/socresonline/3/3/4.html/

Blismas, N. and Dainty, A. (2003) Computer-aided qualitative data analysis: Panacea or paradox?, *Building Research and Information*, 31(6), 455–463.

Bourdon, S. (2002) The integration of qualitative data analysis software in research strategies: resistances and possibilities, *Forum: Qualitative Social Research*, 3(2) http://www.qualitative-research.net/fqs/

Bringer, J.D., Johnston, L.H.J. and Brackenridge, C.H. (2004) Maximising transparency in a doctoral thesis: The complexities of writing about the use of QSR NVIVO within a grounded theory study, *Qualitative Research*, 4(2), 247–265.

Bringer, J.D., Johnston, L.H.J. and Brackenridge, C.H. (2006) Using computer-assisted qualitative data analysis software to develop a grounded theory project, *Field Methods*, 18(3), 1–21.

Carvajal, D. (2002) The artisans tools: Critical issues when teaching and learning CAQDAS, *Forum: Qualitative Social Research*, 3(2). http://www.qualitative-research.net/fqs/

Coffey, A., Holbrook, B. and Atkinson, P. (1996) Qualitative data analysis: Technologies and representation, *Sociological Research Online*, 1(1). http://www.socresonline.org.uk/socresonline/1/1/4.html/

Denzin, N.K. and Lincoln, Y.S. (eds) (2000) *Collecting and Interpreting Qualitative Materials* (2nd Edition), Sage, London.

Dey, I. (1993) *Qualitative Data Analysis: A User-Friendly Guide for Social Scientists*, Routledge, London.

Di Gregorio, S. (2003) Analysis as cycling: Shifting between coding and memoing in using qualitative analysis software, *Strategies in Qualitative Research: Methodological Issues and Practices Using QSR NVivo and NUD*IST*, London, 8th–9th May.

Fielding, N.G. (2002) Automating the ineffable: Qualitative software and the meaning of qualitative research, in May, T. (ed.) *Qualitative Research in Action*, Sage, London, pp. 161–178.

Fielding, N.G. and Lee, R.M. (1991) *Using Computers in Qualitative Research*, Sage, London.

Fielding, N.G. and Lee, R.M. (1998) *Computer Analysis and Qualitative Research*, Sage, London.

Fielding, N.G. and Lee, R.M. (2002) New patterns in the adoption and use of qualitative software, *Field Methods*, 14(2), 197–216.

Fisher, M.D. (1997) *Qualitative Computing: Using Software for Qualitative Data Analysis*, Aldershot, Avebury.

Gilbert, L.S. (1999) *Reflections of qualitative researchers on the uses of qualitative data analysis software: An action theory perspective*, Doctoral Thesis, University of Georgia, Athens, Georgia.

Gilbert, L.S. (2002) Going the distance: Closeness in qualitative data analysis software, *International Journal of Social Research Methodology*, 5(3), 215–228.

Glaser, B.G. and Strauss, A.L. (1967) *The Discovery of Grounded Theory: Strategies of Qualitative Research*, Aldine, Chicago.

Johnston, L. (2006) Software and method: Reflections on teaching and using QSR Nvivo in doctoral research, *International Journal Social Research Methodology*, 9(5), 379–391.

Kelle, U. (1995) *Computer Aided Qualitative Data Analysis, Theory, Methods and Practice*, Sage, London.

Kelle, U. (1997) Theory building in qualitative research and computer programs for the management of textual data, *Sociological Research Online*, 2(2). http://www.socresonline.org.uk/socresonline/2/2/1.html/

Koenig, T. (2004) *Routinizing Frame Analysis through the Use of CAQDAS*, Paper presented at the Bi-annual RC-33 meeting, Amsterdam, 17th–20th August, 2004.

Lee, R.M. and Fielding, N. (1996) Qualitative data analysis: Representations of a technology: A comment on Coffey, Holbrook and Atkinson, *Sociological Research Online*, 1(4). http://www.socresonline.org.uk/socresonline/1/4/1.html/

Lewins, A. and Silver, C. (2007) *Using Software in Qualitative Research: A Step-by-Step Guide*, Sage, London.

Lonkila, M. (1995) Grounded theory as an emerging paradigm for computer-assisted qualitative data analysis, in Kelle, U., Prein, G. and Bird, K. (eds) *Computer-Aided Qualitative Data Analysis: Theory, Methods and Practice*, Sage, London, pp. 41–51.

Miles, M.B. and Huberman, A.M. (1994) *Qualitative Data Analysis: An Expanded Sourcebook* (2nd Edition), Sage, London.

Richards, L. (2002) Qualitative computing: A methods revolution?, *International Journal of Social Research Methodology*, 5(3), 263–276.

Richards, L. (2004) Validity and reliability? Yes! Doing it in software, *Strategies in Qualitative Research*, Durham, UK, 1st–3rd September.

Richards, L. (2005) *Handling Qualitative Data: A Practical Guide*, Sage, London.

Seidel, J.V. (1991) Method and madness in the application of computer technology to qualitative data analysis, in Fielding, N.G. and Lee, R.M. (eds) *Using Computers in Qualitative Research*, Sage, London.

Seidel, J.V. and Kelle, U. (1995) Different functions of coding in the analysis of textual data, in Kelle, U., Prein, G. and Bird, K. (eds) *Computer-aided Qualitative Data Analysis: Theory, Methods and Practice*, Sage, London, pp. 52–61.

Spencer, L., Ritchie, J., Lewis, J. and Dillon, L. (2004) *Quality in Qualitative Evaluation: A Framework for Assessing Research Evidence* (2nd Edition), Cabinet Office, London.

Strauss, A. and Corbin, J. (1998) *Basics of Qualitative Research: Techniques and Procedures for Developing Grounded Theory* (2nd Edition), Sage, London.

Tesch, R. (1990) *Qualitative Research: Analysis Types and Software Tools*, Falmer Press, London.

Welsh, E. (2002) Dealing with data: Using Nvivo in the qualitative data analysis process, *Forum: Qualitative Social Research*, 3(2) http://www.qualitative-research.net/fqs/

Weitzman, E. (2000) Software and qualitative research, in Denzin, N.K. and Lincoln, Y.S. (eds) *Handbook of Qualitative Research* (2nd Edition), Sage, London.

Weitzman, E. (2003) Software and qualitative research, in Denzin, N.K. and Lincoln, Y.S. (eds) *Collecting and Interpreting Qualitative Materials* (2nd Edition), Sage, London, pp. 310–339.

Weitzman, E. and Miles, M. (1995) *Computer Programmes for Qualitative Data Analysis: An Overview*, Sage, London.

Chapter Thirteen
Getting started in quantitative analysis

Chris Leishman

Introduction

This chapter introduces a number of concepts in quantitative analysis. The focus is on inferential statistics and econometric modelling. The purpose of the chapter is to set out an introduction to a range of potentially useful quantitative methods in built environment research rather than to cover every possible aspect of the approaches examined here. Indeed, built environment researchers serious about conducting quantitative research are urged to go further after reading this book and to consider reading more specialised econometric texts such as Green (1997) or Pindyck and Rubinfeld (1997). For those not quite ready to take this next step into the rather complex realm of econometric theory, I set out a more detailed but I hope still understandable version of this chapter in Leishman (2003).

Most researchers utilising quantitative data will use a combination of descriptive and inferential statistics but, in the end, the answers to the research questions set are almost always established with reference to inferential statistics, that is those that rely on sampling theory rather than description. Generally, there is a finely balanced relationship between descriptive and inferential statistics. Preliminary analyses drawing on descriptive statistics are a useful step in conducting more detailed empirical work. For example, when investigating whether variable Y is influenced by variable X then as part of an empirical investigation it would be useful to know the minimum, maximum, median and mean values of these variables. Other measures that are of interest are the variance (or standard deviation) and correlation. Collectively, the measures will indicate the spread of the data and whether there is an association between the two variables. Of course, the key limitation of descriptive statistics is that they do not address questions of cause and effect.

This chapter examines the use and application of quantitative methods in Built Environment research. It is assumed that the research is concerned with the business of testing hypotheses or theorised statements of cause and effect and that the researcher will be working with reference to the laws of probability (sampling theory). After a very brief examination of sampling theory and hypothesis testing, a range of applications are considered ranging from the relatively simple application of parametric testing through simple regression models to more complicated regression models including models of choice, time series and panel models. The latter are examined only in passing since the theoretical and practical issues raised by these approaches require far more comprehensive treatment than possible within a single chapter.

The essence of sampling theory

Key to 'answering' questions in research is the fact that statistical analysis of one kind or another is performed on a sample of data, often collected for the specific purpose of analysis in relation to a given research project. In such projects, statistics are often calculated using the sample of data and the results of these statistical analyses are used to make inferences about the population. The latter, of course, has an important and well-known meaning in statistics and simply means the sum of all observations or cases over which the hypothesis is conjectured to hold.

The case for basing analysis on a sample of data rather than the population is quite simply that it will often be at least impractical, and possibly impossible, as well as almost always inefficient to collect data for a population. A sample is a small set of data drawn from a population. In most cases the researcher will have chosen the data collection methods such that the sample is sufficiently and demonstrably representative of the population in order to allow analysis of the sample to be used to form inferences about the population.

Many, but not all, variables used in quantitative analysis will be Normally distributed (or have a distribution approximating to the Normal distribution). What does this mean? A distribution is an arrangement of the values of a variable in relation to the mean value of the variable. In a standard Normal distribution, the mean is equal to zero and the distribution is symmetrical. Importantly, this means that individual values drawn from the distribution are equally likely to be positive or negative. A distribution can be described by a curve or by a probability density function (pdf). The cumulative probability density function for the Normal distribution takes the following form:

In Figure 13.1, the line $f(x)$ extends from $-\infty$ to $+\infty$ but it should be fairly clear that the probability is almost zero when the standard deviation is -3 and almost one when the standard deviation is $+3$. When the standard deviation is zero, the probability is 0.5. These facts mean that there must be a 50 per cent probability that a randomly drawn value of x will lie within the range $-\infty$ to 0 standard deviations. Equally, there is a 50 per cent probability that the value will lie within the range 0 to ∞ standard deviations from zero. The shape and position of the Normal distribution are determined by the mean and standard deviation (μ and σ). The following Normal distribution (Figure 13.2) has a mean of zero and a standard deviation of 1.

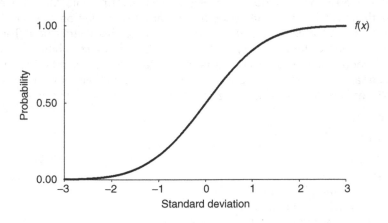

Figure 13.1   Cumulative probability function for the Normal distribution.

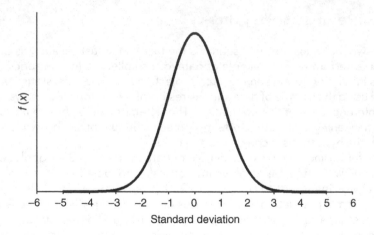

Figure 13.2   Normal distribution with a standard deviation of 1.00.

Sampling theory tells us that the area under the curve between −1.96 and +1.96 standard deviations of the mean is equal to 0.95. To put this another way, if we were to randomly draw a value of x then there is a 95 per cent probability that its value will be between −1.96 and 1.96.

We have now recapped on sufficient facts from sampling theory to remind ourselves how hypothesis testing works. Here is an example. Suppose we are given the value of some unknown variable and asked whether it is likely that this number is, in fact, an observation of the variable x. Now suppose that the mean of the continuous random variable x is 0 and the standard deviation is 1. If the value of the unknown variable is 3 then we can say with more than 97.5 per cent certainty that this is unlikely to be an observation of x.

This logic is at the heart of one of the simplest forms of hypothesis test – the z test (the z score is a closely related concept). If we state the observed value of variable as a 'distance' from an hypothesised population mean then we can weigh up whether the observation is likely to have come from that hypothesised population. For example, if we have a sample whose mean is 50 and standard deviation 10 then we can consider the value of a new variable (suppose it is 75) in relation to that. The z score for the new variable is 75 − 50 = 25 ÷ 10 = 2.5. So, the variable value represents 2.5 standard deviations from the mean of the hypothesised parent sample (population) and it is therefore less than 2.5 per cent likely that the value belongs to this distribution. This simple test is nevertheless potentially very useful and it also has value in helping our understanding of more complex statistical tests. As a rule, the z test can be used when we have a large sample (of at least 30, but preferably much larger) and the data are approximately Normally distributed. The equation for the z statistic is as follows:

$$z = \frac{\bar{x} - u}{s/\sqrt{n}}$$
(13.1)

where

z   z statistic
$\bar{x}$ the observed sample mean
u   the hypothesised population mean
n   the total number of observations in the sample
s   estimated standard deviation of the population

Other common forms of hypothesis test

Although the z test is useful from a demonstrative point of view, the t test is much more commonly used in practice. The simplest way to conceptualise the t distribution is to consider it as a Normal distribution which has been adjusted to take account of the inaccuracies that can arise when dealing with small samples. Statistical tests based on the Normal distribution are not robust with respect to sample size so in many cases the t distribution provides a more robust alternative. The t critical values differ substantially from the Normal distribution when the number of degrees of freedom is small but the difference gradually diminishes as the degrees of freedom increase. When around a thousand degrees of freedom are available the difference between t and Normal critical values becomes negligible.

One application of the t distribution is simply to use the t critical values in place of the z critical values for the purpose of testing an hypothesis concerning a single observation and a conjectured parent distribution. Another common application is the paired t test. This is appropriate when we are constructing a hypothesis about one sample, usually before and after some event or influence. Quite often, the purpose of such a test is to find out whether the event or influence has had a significant effect. For example, we might use a paired t test to compare the best lap times of a group of athletes before and after a change to their exercise regime to test the effect of the change. The formula for calculating the t statistic is as follows:

$$t = \frac{\bar{d} - u}{s/\sqrt{n}} \tag{13.2}$$

where
t t statistic
$\bar{d}$ observed mean difference, that is the sum of $(x_{2i} - x_{1i})$ where x_{2i} is the ith observation of the second sample and x_{1i} is the ith observation of the first sample
u the hypothesised population mean (normally this is equal to zero)
n the total number of observations in the sample
s estimated standard deviation of the population

A particularly favourite form of simple hypothesis test with many researchers keen to integrate qualitative and quantitative methods is the chi-square (pronounced 'kye square') test or χ^2 test. The chi-square statistic is useful for testing hypotheses that draw on categorical data or proportions. The test statistic is calculated as the sum of squared residuals divided by predictions or:

$$\chi^2 = \sum \frac{(Observed - Expected)^2}{Expected} \tag{13.3}$$

The 'expected' proportion is often set as an arbitrarily defined benchmark to allow the test to focus on some second group. The objective is normally to determine whether the second group follows the expected proportion of the first group, the benchmark. For example, suppose we have surveyed a sample of construction professionals ($n = 40$) and our hypothesis is that 'experienced' construction professionals are more likely to believe they require no further training than inexperienced professionals. Suppose 22 of the sample are 'experienced' and 16 expressed this view while 18 of the sample are 'inexperienced' and 10 expressed a similar view. Clearly, the proportions are 72.7 per cent and 55.6 per cent respectively and so they are different. The question is whether this difference is statistically significant. We can use the chi-square statistic. How we compute it depends on how the null hypothesis is set up. For example:

H_0: 50 per cent of each group believe that they require no further training:

$$\chi^2 = \frac{(72.7 - 50)^2}{50} + \frac{(55.6 - 50)^2}{50}$$

$$= 10.933$$

Or

H_0: The same proportion of the experienced group as the inexperienced group will believe they require no further training:

$$\chi^2 = \frac{(72.7 - 55.6)^2}{55.6} + \frac{(55.6 - 55.6)^2}{55.6}$$

$$= 5.259$$

The hypothesis can be formally tested by comparing the chi-square statistic to its critical value (from statistical tables). The sample of data contains two degrees of freedom and so we must look up the chi-square critical value for $2 - 1 = 1$ degree of freedom. This is a common source of confusion in chi-square tests but to put it simply, each respondent in the dataset we have collected can be summarised using two pieces of information: whether they are experienced (or inexperienced) and whether (or not) they expressed the view that they require no further training. Thus, there are two degrees of freedom and not 40 (the sample size)! A short selection of chi-square critical values is reproduced in Table 13.1.

In this case, both of the null hypotheses we defined are rejected. In the first case, we can reject the null that both groups are 50 per cent likely to express the view; in the second case, we reject the null that the same proportion of the experienced group express the same view as the inexperienced group. However, notice that the null cannot be rejected at the 1 per cent significance level in the latter case.

Inference and causality – basic regression models

Postgraduate students involved in quantitative built environment research are likely to consider moving further beyond descriptive statistics and hypothesis testing to the wider area of econometric modelling. The step from descriptive to inferential statistics (including hypothesis testing) allows the researcher to move beyond discussion of

d.f.	Significance level	
	5%	1%
1	3.841	6.635
2	5.991	9.210
3	7.815	11.345
4	9.488	13.277
5	11.07	15.086

Table 13.1 Chi-square critical values.

results and begin making more scientific assertions and conclusions. One of the key limiting factors in hypothesis testing is the relative simplicity of the process: normally the researcher sets up a null and an alternative hypothesis. The hypothesis is typically along the lines of 'X causes Y'. The testing process will lead the researcher to be able to reject the null hypothesis, or fail to reject it. Although it is very useful to the researcher to be able to make a scientific conclusion such as this, the process is still very restrictive. For example, if there are many potential influences or determinants of Y (and not just a single variable, X) then the researcher would need to carry out a series of hypothesis tests, each involving the test of whether a different X variable is statistically related to Y. Although the laborious nature of this is unattractive in its own right, there is a more far-reaching problem which is that the researcher is implicitly assuming that no relationship exists between the Xs when following such an approach. In addition, the restrictiveness of the hypothesis testing approach is such that the process will not yield any information on the respective strengths of association between the various X variables and Y. It is not valid, for example, to draw any judgement about the relative size of the test statistic when carrying out a series of tests.

These restrictions often lead researchers to employ more flexible and powerful statistical methods. In the remainder of this chapter, we will briefly examine regression analysis or econometric modelling. The examination is brief because the field is very well developed and the term 'econometric modelling' describes a wide-ranging field encompassing spatial, time series, cross-sectional and panel approaches. Econometrics is an established academic discipline in its own right and this chapter goes no further than to consider the basic logic of econometric approaches. Chapter 5 deals with approaches to economic modelling and analysis and their applications to the built environment.

Perhaps the most significant conceptual difference between descriptive and inferential statistical analysis is the fact that researchers using econometric approaches are seeking to establish models. A 'model' has a particular and significant meaning in academic research but, clearly, a model is a simplification of, or abstraction from, reality that retains the important aspects or relationships involved. Ideally, statistical models should be capable of satisfactorily *explaining* how our chosen aspect of reality works. Second, the model should be capable of satisfactory *prediction*. The regression model, as a concept, therefore holds considerable appeal for the researcher. A good econometric model will yield information about which variables determine the variable of interest (referred to as the dependent variable). After estimation, an econometric model can be used to predict or model the data. Given new values of the independent or explanatory variables, it should also be possible to use the estimated model to forecast unknown values of the dependent variable. However, econometric models require much more extensive testing and scrutiny than simple hypothesis tests and, as a casual observation, it is worth noting that there is a great deal of literature focusing on the identification of problems caused (and solutions) when econometric models go wrong.

The most complex of econometric models still rests on the initial assumption that a line of best fit may be used to approximately describe the relationship between two variables, X and Y, or the independent and dependent variable:

$$Y_i = a + bX_i + u_i \tag{13.4}$$

where
Y_i the estimated value of the dependent variable for the ith observation
a constant (the same for all observations)
b the slope or gradient of the regression line (the same for all observations)

X_i the value of the independent variable for the ith observation
u_i the error term or residual for the ith observation

The error or disturbance term (u_i) is present to ensure that (hopefully) small differences between the explained or predicted values of the dependent variable and the observed values can be ascribed somewhere. In other words, the error term is the variable that takes on any element of the dependent variable not explained by the model.

In fact, the error term is actually the key to determining the line of best fit because the regression line is defined as the line that minimises the amount of unexplained variation in the dependent variable (Y). Total variation in the dependent variable Y is defined as follows:

$$\sum (Y_i - \overline{Y})^2 \qquad (13.5)$$

This is known as the total sum of squares (*TSS*) and it measures the dispersion of the observed values for Y about the mean value of Y. The total explained variation in the dependent variable Y is known as the regression sum of squares (*RSS*) because it measures the dispersion of the estimated values for Y about $\overline{Y}$. It is defined as

$$\sum (\hat{Y}_i - \overline{Y})^2 \qquad (13.6)$$

The unexplained variation in the dependent variable Y is known as the error sum of squares (*ESS*) because it measures the dispersion of the observed values for Y about the estimated values for Y. Its definition is

$$\sum (Y_i - \hat{Y}_i)^2 \qquad (13.7)$$

Therefore, in summary, the solution for the regression line (the estimated values for *a* and *b*) are determined by fitting a line to a scatterplot of data such that the *ESS* is minimised and no other position or gradient would reduce it further. Since we have three measures of dispersion in the data (total variation or *TSS*, explained variation or *RSS* and unexplained variation or *ESS*), we can also construct a statistic measuring the ration of explained to total variation. This is known as the R squared statistic and is defined as follows:

$$R^2 = \frac{RSS}{TSS} \qquad (13.8)$$

When data points lie close to the regression line then the variance explained by the line will be close to the total variance in the data and the R squared statistic will be close to 1. When the data are widely dispersed about the line then the R squared statistic will lie close to zero.

Multiple regression models

Multiple regression analysis differs from simple regression in that more than one explanatory or independent variable is used to explain and predict the dependent variable, that is a combination of independent variables jointly determines the dependent variable. Most applications of econometrics will use multiple, rather than simple, regression analysis. In multiple regression analysis it is more usual to use the adjusted R^2 rather than the R^2 statistic. This is because the R^2 statistic cannot reduce with the inclusion of more explanatory variables. The adjusted R^2 is simply the R^2 statistic weighted to account for the number of parameters estimated:

$$\overline{R}^2 = 1 - (1 - R^2)\frac{N-1}{N-k}$$ (13.9)

where
N number of observations
k number of independent variables

The adjusted R^2 statistic may either increase or decrease with the addition of a 'new' explanatory variable so researchers often use the statistic in an iterative manner in order to gauge (crudely) whether or not a model specification has been optimised.

Researchers using multiple regression methods also use the F statistic to test whether the overall regression equation is statistically significant. The null hypothesis is that all of the equation parameters are equal to zero. If we reject this hypothesis then we fail to reject the alternative hypothesis that the equation is of some use in explaining and predicting the dependent variable. The F statistic is calculated as follows:

$$F_{k-1,n-k} = \frac{R^2/(k-1)}{(1-R^2)/(n-k)}$$ (13.10)

The critical value for the test statistic is found in the F distribution with degrees of freedom equal to the number of parameters less one (numerator) and the number of observations less the number of parameters (denominator).

In multiple regression analysis, researchers often use the R squared, adjusted R squared and F statistic together, and in an iterative process, to determine whether the model has any value overall. The value of including individual variables in the regression equation is determined primarily in relation to the t statistic. Unlike the R square, adjusted R square and F statistics, which test the equation overall, standard regression output will yield one t statistic for each variable included in the model.

The t statistic is a test statistic, that is it is used to test an hypothesis (that the true value of the parameter is zero). For each parameter, the test statistic is the estimated parameter (coefficient) less the hypothesised 'true' population value (zero), all expressed as a proportion of the standard deviation of the parameter estimate:

$$t = \frac{\hat{b} - b}{s_b}$$

The critical value depends on the number of degrees of freedom since we are working with the t distribution but, as a rough rule of thumb, we might expect statistical significance when the t statistic is greater than 2 or smaller than −2. Such a result implies that the estimated parameter of the variable concerned is more than 2 standard deviations distant from zero, so we can reject the null hypothesis that the coefficient is non-zero by chance and might be zero if we were to collect a different sample of data and estimate the parameters again.

The use and interpretation of t statistics is more important in multiple than simple regression analysis. In the latter, it is obvious that the independent variable is statistically significant if the model has a good fit (high R^2 and F statistics). In multiple regression analysis it is possible to specify and estimate a model which has a good fit but in which not all (or even relatively few) of the explanatory variables are significant. Examining the t statistics permits the identification of the explanatory variables that are significant. Variables that are not significant are normally dropped and the analysis is repeated, a process which may help to identify a model specification with higher explanatory power.

As some of the preceding discussion suggests, researchers often work in an iterative way, particularly when using econometric methods. The appeal lies in the ability to fine-tune a model which already shows some promise and results not far removed from prior expectations. In other words, there is a fine line between good practice in model optimisation and data mining! The latter is generally regarded as poor practice in the application of econometric models, though such methods may have a place elsewhere. The duties of a researcher in ensuring that a sound econometric model has been estimated extend far beyond care in model refinement. In particular, econometric models are valid only when an important set of assumptions and rules has been preserved. The assumptions are as follows:

The error term is assumed to have a zero mean and is Normally distributed. This means that there will be as many positive as negative values while the majority of the residuals will be distributed closely around zero. The errors should have constant variance; independent variables are assumed to be uncorrelated with the residuals and, finally, there is assumed to be no exact relationship between independent variables.

When the errors are not Normally distributed, this casts doubt on the standard errors and increases the risk of wrongly rejecting one or more null hypotheses regarding the significance of the parameter estimates. In other words, the estimated model may appear better than it really is!

In some cases, violation of the regression assumptions is suggestive of model misspecification (or incorrect functional form). For example, suppose that the true relationship is a quadratic between Y, X and X^2, that is $Y = a + b_1 X$ and $b_2 X^2$. Now suppose that we regress Y on X and fail to observe X^2 or include it in the model. What is likely to happen? Since there is a relationship between X and Y, we should obtain adjusted R square, F and t statistics suggestive that the model has some explanatory power. However, if we were to examine the residuals, particularly if we ordered them with respect to X, we would discover a relationship within the series. This is simply because the quadratic part of the function was not accounted for in the regression equation, hence this element of the relationship between X and Y remains in the residuals. A symptom of heteroscedasticity is incorrect estimates of the standard errors. If the residuals are heteroscedastic then we can no longer rely on our test statistics. Figure 13.3 summarises one possible visual pattern of heteroscedastic residuals.

Testing for heteroscedasticity is relatively straightforward and most statistics software packages offer fairly accessible options. Common tests include the

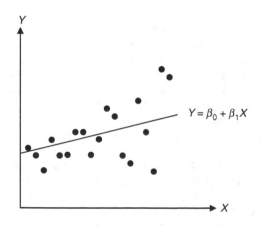

Figure 13.3   An example of heteroscedastic residuals.

Goldfeld-Quandt, Breusch-Pagan and White tests (see Green, 1997; Pindyck and Rubinfeld, 1997 for a full discussion). The Goldfeld-Quandt test is one of the easiest to conceptualise. The data are ordered with respect to one of the independent variables and separate regressions are run for low and high values. Sometimes a small number of middle observations may be omitted altogether. The ratio of the error sum of squares of the second regression (involving high values of the independent ordering variable) to the first is tested against the F distribution. When the ratio is higher than the critical value the null hypothesis of homoscedasticity is rejected (heteroscedasticity cannot be ruled out).

Correlation between independent variables is a common failing of econometric models, particularly cross-sectional models. An exact linear relationship between explanatory variables is referred to as perfect collinearity and will make parameter estimation impossible. However, it is high, rather than perfect, collinearity between independent variables that causes more empirical problems in practice. When this occurs, it is possible to estimate the regression parameters but the results may be misleading. This phenomenon, known as multicollinearity, may be difficult to detect. A common set of symptoms is the joint presence of high R^2 and F statistics but low t statistics for some or all of the explanatory variables (an illogical result). Another common indication is volatile variable parameters with respect to model specification. In other words, when some variable coefficients change markedly depending on which other variables are also included in the model, then there is some suggestion of multicollinearity.

It is common practice to produce a matrix of correlation coefficients prior to (or even just after) estimating the regression parameters. The correlation matrix may help to identify pairs of explanatory variables that are highly correlated. Another common multicollinearity solution is to include collinearity diagnostics in model estimation output and use this to guide final model specification.

Concluding remarks

The purpose of this chapter is not to provide a detailed manual for undertaking quantitative analyses but to offer some insights and suggestions for researchers considering a quantitative approach. The chapter has covered sampling theory, hypothesis testing and basic concepts in ordinary least-squares (OLS) regression. It should provide the reader with sufficient background to begin exploratory analysis coupled with further reading designed to provide a deeper understanding of some of the issues explored.

Of course, the other obvious difficulty faced by many researchers considering quantitative methods is access to, or choice of, software. Recommending appropriate software is always difficult because there exists a considerable range of statistical and econometric software and the advantages and disadvantages of each can be subtle. However, researchers with a very limited background in statistics might want to consider a package such as Excel to begin with. This is a much undervalued software package when undertaking statistical and econometric analysis. Yet, it has useful capabilities in hypothesis testing, descriptive statistics and regression estimation. Meanwhile, SPSS is the tried and tested workhorse application for many students and professional researchers and no discussion of statistics software would be complete without a mention of it. Almost all universities, and many public and private research organisations, own site licences for SPSS and for many readers it will be the most accessible software. Finally, other aspects of econometric modelling are examined in Chapter 5 and it should be noted that neither Excel nor SPSS are specifically designed to handle time series, spatial or panel econometric approaches. These, more specialised, applications are well beyond the scope of this book and interested readers are referred to Greene (1997) or Pindyck and Rubinfeld (1997) for a thorough discussion.

References

Greene, W.H. (1997) *Econometric Analysis*, Prentice-Hall, New Jersey.

Leishman, C. (2003) *Real Estate Market Research and Analysis*, Palgrave Macmillan, Basingstoke.

Pindyck, R.S. and Rubinfeld, D.L. (1997) *Econometric Models and Economic Forecasts*, McGraw-Hill, Boston.

Chapter Fourteen

Artificial neural network modelling techniques for applied civil and construction engineering research

Abdelhalim Boussabaine and Richard Kirkham

Introduction

Drawing upon the complex biological structures of the human brain and central nervous system, Artificial Neural Networks (ANNs), referred to sometimes as simply neural networks although this is not technically accurate, are systems that are embedded with the ability to learn. By replicating (in some measure) the methods that biological systems induce when making decisions, solving problems or understanding information, ANNs can offer a powerful modelling approach to a vast array of research problems. The applications of this technique pervade a wide spectrum of academic disciplines. Within construction and civil engineering specifically, the early 1980s were perhaps a 'watershed' in the development of ANNs. Consequently, a number of applications now exist for 'non-computer' specialists, which has empowered those equipped with a relatively modest knowledge of the theory behind ANNs to tackle complicated problems in a range of specialist areas. This chapter aims to provide the reader with a general introduction to ANNs, some background introduction to methods and methodology, a review of the current research landscape and an evaluation of the most significant applications of ANN techniques in the field. The chapter also briefly explores Neuro-fuzzy systems. These are effectively a sub-set of ANN theory and combine the knowledge representation power of fuzzy logic (fuzzy logic in the broad sense deals with approximated reasoning) with the learning potential of ANNs. The overall content of the chapter is positioned within the construction and built environment research landscape. It therefore attempts to offer a cutting edge examination of current ANN theory and application.

First concepts

ANN modelling differs from statistical modelling in the sense that in regression models (say), the approximated function is assumed (by the researcher) and the regression coefficients are determined by iterative calculation. In contrast, the ANN is asked to approximate the unknown function that maps the input space to output. In other words, a function is being asked to approximate another function. The model is said to be able to solve the problem if it is to learn to approximate the function to an arbitrary accuracy. This process is demonstrated in Figure 14.1.

Consequently, traditional models generally lack the ability to learn inherently, generalise solutions and respond adequately to highly correlated, incomplete or

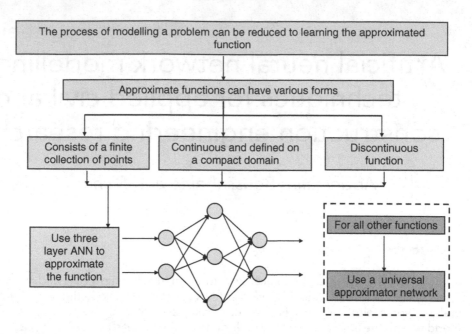

The process of modelling a problem can be reduced to learning the approximated function

Approximate functions can have various forms

| Consists of a finite collection of points | Continuous and defined on a compact domain | Discontinuous function |

Use three layer ANN to approximate the function

For all other functions

Use a universal approximator network

Figure 14.1   Approximation of functions by ANN models.

nebulous data. For this type of environment, neural models are generally superior (Boussabaine *et al.*, 1999). The critical distinction over simple mathematical and statistical models lies in *adaptivity*. Neural systems can automatically adjust their weights to optimise their behaviour as decision makers and predictors. Thus, self-optimisation allows the neural system to effectively 'design itself'. In the construction management discipline, for example, ANNs can be seen as sub-systems which exploit expert-given rules or statistical inference techniques. Such systems, in turn, will be able to provide decision support for experts, increase the efficacy of the decision-making process at complex levels and assist in the training of inexperienced personnel with risk management and scenario planning. A neural system can learn (be trained) on a set of input and output data belonging to a particular problem. If new data of the same problem, but not in the training set, are presented to the system, the network can use the 'learned' data to predict outcomes without any specific programming relating to the category of events involved. The fields of application of ANNs have increased dramatically in the past few years. A large variety of possible ANN applications now exist for non-computer specialists. Therefore, with only a very modest knowledge of the theory behind ANNs, it is possible to tackle complicated problems in a researcher's own area of speciality. Within construction and civil engineering, the applications are widespread and this will be covered later.

A useful ANN decision support system must be robust, easy to use and should enhance the decision-making process. An ANN that simply models the decision-making behaviour of the user is likely to be of limited use. This review will concentrate on the neural techniques and concepts useful in construction management and help researchers to identify opportunities where this technology is applicable in assisting decision makers.

Obviously, one of the key benefits of ANNs is the ability to be used as a function approximation mechanism which learns from sample data. There are however key issues that the researcher should address before employing these methods. Whilst the use of AI software now permits relatively straightforward implementation of ANNs to problem statements, researchers should be aware of issues such as choice of model, learning algorithm and robustness. (Other parts of this book explore aspects of these themes.) In addition, this chapter provides guidance on the development of successful applications in construction management and civil engineering practice.

System dynamics

The computational dynamics and complexity of ANNs are radically different to conventional analytic methods (i.e. regression analysis, linear programming, etc.). By simulating the biological structure of the human central nervous system (CNS), ANNs function by learning from experience and generalising from previous examples to new one's and abstracting essential characteristics from inputs containing irrelevant data. Network components with names such as neurons (sometime referred to as cells, units or nodes) and synaptic transmissions with weight factors are used to mimic the nervous system (analogous to synaptic connections in the nervous system) in a way which allows signals to travel through the network in parallel (ANNs are often referred to as a parallel distributed network) as well as in serial.

Whilst ANNs intrinsically exhibit a number of virtues in common with the human brain, this resemblance is arguably a superficial one. Haykin (1994) (in Ok and Sinha, 2006) contends that 'in practice nowadays it can be said that the ANN only represents the brain at the most elementary level of process, although the ANN has retained as primary features two characteristics of the brain: the ability to "learn" and to generalise from limited information'. Haykin's pragmatism is also evident in Boussabaine's (1996) review of ANN applications (in construction and civil engineering research), which presents a most plausible hypothesis that the likelihood of these networks ever being able to truly replicate the functions and complexity of the human brain is extremely remote. Flood (2006), in his analysis of ANN developments, (this is discussed in a later section) uses the number of primary processing units that can be usefully employed in an application as the measure of complexity, asserting that

> we can compare today's general purpose digital computer to the brain of a rabbit (comprising in the order of 10^9 neurons), while ANNs have progressed no further than the brain of the humble nematode (comprising just 302 neurons).

A number of models (and variants) have been proposed in the literature. Recent summaries of the significant contributions can be elicited from Wang and Elhag (2007) and Kumar (2005) as well as earlier contributions to the field crystallised in Boussabaine (1996, 2001), Lippman (1987) and Hush and Horne (1993). It is perhaps interesting to note at this juncture that the applications of ANN models have been amply demonstrated in the 'softer' scientific fields such as human resource management (Huang et al., 2006), prediction of construction litigation and dispute resolution outcomes (Arditi and Pulket, 2005) and productivity and plant management (Chao and Skibniewski, 1994) as well as the 'hard' physical sciences such as transient heat-flow modelling in buildings (Flood et al., 2007) and modal seismic control of building frames (Rao and Datta, 2006).

Historically, the most popular models in the broadest applications are *non-linear multilayered networks*. Emerging from the single-neuron linear ANN approach of the 1960s, original variants were composed of two layers of computational neurons: input and output neurons. Training of the models is relatively straightforward and these have found widespread commercial application as a consequence (Fausett, 1994). The relative simplicity of the approach does, however, yield only a limited number of applications – this led to the introduction of *non-linear multi-element* networks and algorithms for calculating and correcting errors within the network performance, such as the now commonplace *back-propagation algorithm*.

The development and theory of ANNs from the introductory level to more advanced strategies is described in a number of papers. Barron and Barron (1989) and Levin *et al.* (1990) provide a statistical interpretation of training methods in ANNs. An examination of the feed-forward network (these are networks that do not exhibit feedback (loop) properties) and the back-propagation models (where the transfer function used by the networks is differentiable and consists of a feed-forward phase and a backward phase to allow any error between the observed and requested values in the output layer to be modified by adjustment to the weights) can also be found in Kamarthi *et al.* (1992). Kohonen (1988) reported interesting and useful results from his research on self-organising feature maps used for pattern recognition and signal processing. It has been proven that problems which involve complex non-linear relationships could be better solved by ANNs than by conventional methods (Rumelhart *et al.*, 1994). ANNs are suited to such problems because of their proprietary adaptivity; that is, non-linear activation functions (Flood and Kartam, 1994a, 1994b). Adaptivity allows the ANN to perform well even when the environment or the system being modelled varies chrono-logically. ANNs exhibit distinct advantages over traditional methods of modelling in situations where the process to be modelled is complex to the extent that it cannot be explicitly represented in conventional mathematical or statistical terms or that explicit formation causes loss of sensitivity due to over-simplification.

Network structure and nomenclature

Most ANNs consist of a familiar composition of inputs, outputs and a number of 'hidden layers' which are comprised of artificial neurons (these can often be referred to as processing elements (PE), processing units (PU) or artificial neurodes (AN)). These are connected together to form a network – hence the term 'ANN'. The neurons are effectively the vehicle for receiving one or a number of inputs and through a *transfer function*, producing an output. Transfer functions dictate the behaviour of the ANN. This behaviour thus depends on both the weights assigned and the input–output function that is specified for the neurons. There are a number of transfer functions available. Stergiou and Siganos (2007) classified these into *linear* (the output activity is proportional to the total weighted output), *threshold* (the output is set at one of two levels, depending on whether the total input is greater than or less than some threshold value) or *sigmoid* (the output varies continuously but not linearly as the input changes). For the latter transfer function, which is probably the most popular, sigmoid functions 'better' exhibit the nature of biological neurones. Although, of course, they still lack considerable resemblance to a real biological network (like the brain!). Therefore, to ensure that an ANN performs a specific task, the desideratum for neuron connectivity must be established, and furthermore, the weights on the connections should be appropriately selected. Since the connections determine whether it is possible for one unit to influence another, then the importance of weight specificity (the strength of this influence) cannot be underestimated.

In summary, whilst the system architectures can offer a diverse range of options in terms of topology and mode of operation, seven major features can be observed (Lippman, 1987; Hall, 1992; Hush and Horne, 1993) in Boussabaine (1996):

(1)   A set of processing neurons.
(2)   A state of activation for each neuron.

(3) A pattern of connectivity among the neurons or topology of the network.
(4) A method to propagate the activities of the neurons through the network.
(5) An activation rule to update the activities of each node.
(6) An external environment that provides information to the network and interacts with it.
(7) A learning method to modify the pattern of connectivity by using information provided by the external environment.

Figure 14.2 illustrates a typical ANN with three layers. These consist of a number of nodes with each of the nodes in one layer linked to each node in the next layer. The communication with the external system occurs through the nodes of the input and output layers. The hidden layer provides a critical computational ability to the system (sometimes this is referred to as the 'black box'). Figure 14.3 shows a typical neuron structure where inputs from one or more previous neurons are weighted

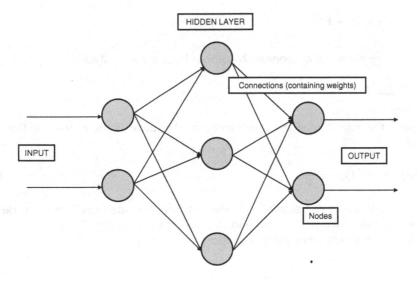

Figure 14.2   Typical three-layer artificial neural network.

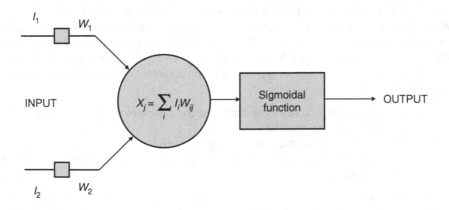

Figure 14.3   Typical activities at the neuron.

and summed. This results in an output value which is passed on to connected neurons in the successive layer.

The general notation for the process is summarised in Wang and Elhag (2007), where each neuron in the hidden layer receives the activation signal (which is the weighted sum of all the inputs entering the neuron)

$$X_j = \sum_i I_i W_{ij} \tag{14.1}$$

To produce an output through the transfer function. Previously, some typical transfer functions are identified, here Wang and Elhag use the sigmoidal transfer function

$$h_j = f(x_j) = \frac{1}{1 + e^{-x_j}} \tag{14.2}$$

With the properties

$$\frac{df}{dx_j} = f(x_j) \left[1 - f(x_j)\right] \tag{14.3}$$

Activation signals to the neurons in the output layer are executed

$$Y_k = \sum_j h_j W_{jk}$$

And finally, the activation signals are transformed to provide the outputs of the neural network:

$$Output_k = f(y_k) = \frac{1}{1 + e^{-Y_k}} \tag{14.4}$$

This is the briefest of explanations of how the network operates. For more detailed coverage of the concepts covered here, the suggested reading list at the end of this chapter provides recommendations of appropriate texts.

System architecture design

The development process of system architecture in ANN models is illustrated in Figure 14.4. The process starts with the evaluation and assessment of the suitability of AI tools to solve the problem under investigation. The criteria for evaluation can be found in Boussabaine (1996) and Masters (1993). Having established the suitability of AI methods, the procedures for data collection and preparation should be established. This aspect of the modelling process is important since it is at this juncture, the modeller is able to:

- Learn more about the nature of the data and the problem under investigation.
- To solve problems within the data.
- To change the structure of data (levels of granularity).
- To extract meaningful knowledge and analysis of the qualitative variables and quantitative variables.

The next stage normally involves prototyping, through definition of input and output variables and the types of decision that the ANN system will make. The expected outcome of this stage is an approximated function that maps out the input to the

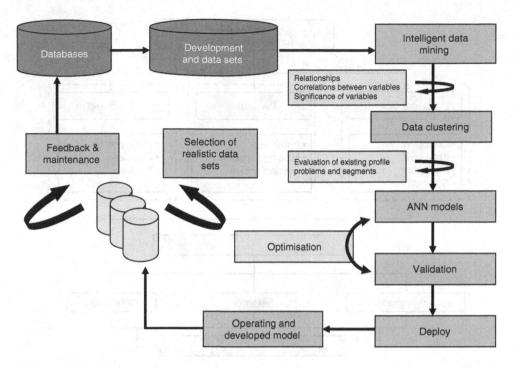

Figure 14.4 The ANN modelling life cycle.

output. This system is then optimised, tested, validated and deployed in a practical setting once performance has attained a satisfactory level. Necessary feedback is then established for future maintenance and development. In summary then, the key issues facing the modeller are:

- Data acquisition, analysis and problem representation.
- Clustering.
- Architecture determination.
- Learning process determination.
- Training of the network.
- Testing of the trained network for generalisation evaluation.

There are no hard and fast rules for defining an ANN. However, the following criteria are suggested as a road map to system design. This section deals with the initial decisions that are taken for developing ANN models, optimisation of generated models, testing and validation. Again, this is intended only as a brief examination of the salient points – the reader should consult the references for more comprehensive coverage.

Data acquisition, analysis and problem representation

Any data collated should be transformed from the raw state into an appropriate form for model generation. (This may not always be necessary, however.) The process of data pre-processing is illustrated in Figure 14.5. The process can involve linguistic variables transformation to numerical and statistical information about data, data cleaning, missing data analysis, replacing missing data via local means and extreme value/outlier analysis. Other preparation to remove any unwanted bias

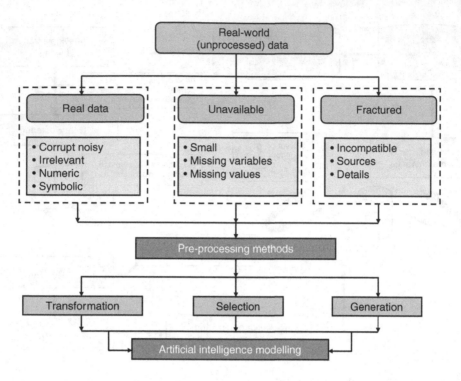

Figure 14.5 The process of data pre-processing.

which data may contain should also be addressed. It may also be the case that normalisation of data should be carried out (i.e. all the data needs to be transformed into a common scale). This avoids weighting being attributed to different features simply on the basis of the scales involved. Statistical analysis may also be required to discover any interdependency between variables. The use of a correlation matrix is often appropriate at this stage. The outcome should be a robust, accurate and normalised database ready for modelling by any of the techniques described here. Data representation and selection of training have a vital role in the performance and convergence of ANN models. The following processes must be carried out on the data sets:

- Analysis of the data sets for the identification of the variables that is important to the process.
- Optimal size of the training and testing data sets.
- Optimal input variables.
- Normalisation of data sets.
- Discovery of errors in data sets.

Clustering of data

Clustering (and dimensionality reduction) is often used in neural network modelling during data pre-processing – it is the process of organisation of data into subsets (sometimes also referred to as partitioning) based on evidence of some homogeneity. This allows grouping of data in each subset which share some characteristics. Elhag and Boussabaine (2001) describe the use a particular type of ANN known as a Kohonen

Self-Organising Map (SOM or SOFM, where feature is used) for data pre-processing and clustering. SOMs are unsupervised learning networks – the network learns pattern recognition automatically. This is as opposed to supervised learning, where a network is trained by presenting an input pattern together with a corresponding output pattern. SOM also provides the opportunity to aid in representation of data and addresses the problem of variability and interdependence between attributes. These neural networks attempt to learn a topological map from an N-dimensional input space to a two-dimensional feature space. Topological means that if two vectors are close together in the input space, their mapped representations will tend to be close together in the two-dimensional space.

Figure 14.6a illustrates the typical map of a 3 × 2 two-dimensional SOM and is an orderly mapping of a high-dimensional distribution of data onto a regular low-dimensional grid. The results produced by SOM algorithms can be considered to be similarity diagrams of data and their clusters.

After training the nodes on data, the output map represents characteristic classes of data sets with similar patterns. An example of an output class is shown in Figure 14.6b, which indicates the strength of the similarity of the patterns within the group represented by the neuron. A strong grouping will have all the patterns close to the centre. A weak group will have patterns widely distributed away from the centre. A weak grouping may indicate the need for more training or for more neurons to be added to the network to allow more groups to be created during training, thus allowing the similarities to be better represented. The closer the pattern to the centre, the stronger the

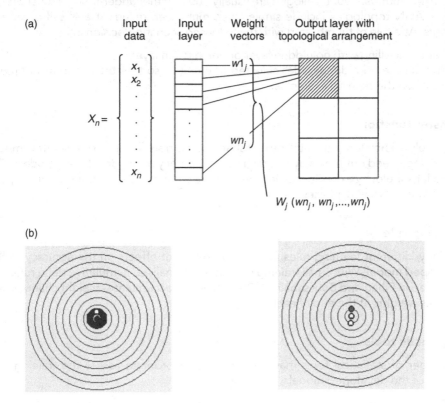

Figure 14.6 (a) Dimensional SOM network; (b) result of clustering.

similarity between the pattern and the centre of the group represented by the neuron. The further from the centre, the weaker the similarity between the pattern and the group. The architecture of the SOM networks consists of:

- An input layer. It includes the nodes representing all the input variables.
- A two-dimensional Kohonen layer. The nodes are arranged in a discrete rectangular grid. At any given iteration, Kohonen learning takes place for the winning node and a set of its spatial neighbours. It is this mechanism of neighbourhood learning that preserves closeness in the mapping.

Here, it is important for the modeller to use the following criteria to select the best data subsets for creating highly accurate ANN models:

- Small coefficient of variation.
- Strong correlation between input and output.
- Dissimilarity with other clusters.
- Significant number of cases (samples) to allow for two sets of data (training and testing).

Topology

There is no formal method to derive a network configuration for a given prediction problem. Therefore, the problem of finding a sensible topology and a good set of parameters is carried out by optimisation and adjustment of the developed model parameters. Poor topologies are usually due to either under-fitting or over-fitting. Typical ANN topologies include single layer, multilayer, recurrent and self-organised variants. As a rule of thumb, start with the following scenario decisions:

- Start modelling with no hidden layer or one hidden layer.
- The number of hidden nodes selected at the start so that total weights are much less than the training examples.

Transfer function

It is widely acknowledged that the exact shape of the transfer functions has little impact on training speed but they may affect ultimate accuracy of the developed models. The modeller should, by experimentation, investigate a number of functions at training and also during modelling phases.

Learning rule

Similarly, different learning rules may exhibit different effects on the accuracy of the developed model. With small training date sets, normalised cumulative delta rules are proposed in Elhag and Boussabaine (2001). This rule accumulates weight changes and updates weights at the end of each epoch.

Diagnostic tools

A number of statistical tools may be deployed in order to assess the training and testing of the developed models, including:

- Root mean squared error for output layer (RMSE).
- Correlation.

- Confusion matrix for each output.
- Histogram analysis of weights for each layer.
- Input contribution.
- Mean Absolute Percentage Error (MAPE).

RMSE and MAPE are generally the most commonly used and are given by (14.5) and (14.6) respectively

$$\text{RMSE} = \sqrt{\frac{\sum_{i=1}^{n}(\hat{x}_i - x_i)^2}{n}} \tag{14.5}$$

$$\text{MAPE} = \left(\sum_{i=1}^{n}\frac{|x_i - \hat{x}_i|}{x_i} * 100\%\right)\bigg/n \tag{14.6}$$

Optimisation

The best optimal network model should consist of a combination of the following characteristics:

- The variance should be explained.
- Should have the minimum number of input variables to reduce complexity.

Learning rate

Earlier in the chapter, the two principal methods of learning were identified, supervised (such that (x, y), $x \in X$, $y \in Y$) and unsupervised, the latter of course being an example of self-organisation given that the desired outputs are not specified. The learning rate can be optimised by using the following procedures:

- The network is initialised and trained for a number of iterations depending on the size of the training set.
- Simultaneously:
  - o The learning rate is lowered so that the RMS error plot is smooth.
  - o The relative magnitude of the learning rates is changed so that the weight histograms for each layer spread at the same rates.

Epoch size

An epoch is the delivery of the training set (data) to the neural network (this is sometimes also referred to as an iteration). The epoch size can be optimised by using the following procedures:

- Establish an initial epoch size. Demir and Ucar (2003) suggest that the iterations (epoch sizes) are task dependent and should be determined by a trial-and-error procedure.
- Initialise the network and train it for a number of iterations depending on the size of the training set.
- Test the network and record the R-coefficient for each output.
- Repeat the above for a variety of epoch sizes.
- Plot out the R-coefficients and select the epoch which produces the highest value.

Hidden layer and node size

The hidden layer and node sizes are optimised by using the following procedures:

- Start with a minimum number of hidden nodes and add more as the training process proceeds.
- Start with the maximum number of hidden units and prune out marginal one's .
- Number of hidden nodes = (inputs + outputs)/2.
- Use a number of nodes such that the number of weights as a product of 10 is smaller than the number of training cases (to avoid over-fitting).
- Trial numerous networks with different frequencies of nodes, estimate error for each one, and select the network with the minimum estimated generalisation error.

Testing

It is very important to recognise that testing is as important as the training of the model. Testing data should not be used as part of the training procedure in any way whatsoever. The testing data set should also be selected randomly from the initial experimental data.

Validation

A side-effect of the 'black box' nature of ANNs is that the system does not inherently provide an obvious explanation of how a problem is solved. In some respects, this is not so surprising given that artificial neural networks are not unlike humans. Both express opinions that they cannot easily explain! ANNs are able to pinpoint certain factors, that were thought to have been irrelevant or which conflict with traditional theories, as important for decision making. This aspect can be extremely frustrating due to the fact that there is no way to determine whether the network has incorrectly identified these factors or if, by chance, traditionally accepted methods are wrong. There is also no assurance that the network will train to the best configuration possible. Even if the neural network is functioning correctly, it can still be prone to errors. However, even so, all networks suffer from limitations in their ability to learn and to recall. The importance of the degree of accuracy must be assessed and then it must be decided if it is worthwhile to use neural networks. The output from the system is valid only if it approximates the function that is the subject of modelling. Reliability of the developed model is measured by the consistency of the results produced by the model. Accuracy is a measure of the fitness of the output from the model and these criteria can be measured using MAPE. Confidence reflects the nature of the data used to develop the model rather than the model itself. Sensitivity is more related to the results than the method of modelling. To further validate accuracy of the models, it may be appropriate to statistically compare the performance of traditional mathematical models when subjected to identical data input.

Recent advances in construction and civil engineering research

Anecdotal evidence suggests that the engineering disciplines continue to explore the applications of ANNs to practical solutions, and to a lesser extent in the traditional construction-related disciplines as well. Flood's recent editorial to the *ASCE Journal of Computing in Civil Engineering* (Flood, 2006) does, however, cast a fascinating reflection on the theoretical advancement of ANNs and associated fields including optimisation. His analysis of submissions to the journal reveals that, since 1995, 12 per cent of papers published include the term 'neural' in the primary title. Flood reinforces the analysis with data from the ISI Web of Science, which indicates that the top 5 most cited papers (in that journal at 2006) were concerned either with ANNs or optimisation procedures (specifically, Genetic Algorithms (GAs)). Nevertheless, Flood's hypothesis is that since the 'watershed', the literature had focused more on application of existing neural theories (based on simple vector mapping techniques) rather than developing neural systems to better emulate the complex cognitive process that the methodology should aspire to. In summary, whilst the extolled virtues of ANNs are well versed, informed comment suggests that a great deal of further research work is required on theoretical development rather than application.

Neuro-fuzzy modelling

Neuro-fuzzy is a combination of the explicit knowledge representation of fuzzy logic with the learning capabilities of ANNs. Neuro-fuzzy modelling involves the extraction of rules from a typical data set, and the training of these rules to identify the strength of any pattern within the data set. The system creates membership functions from which linguistic rules can be derived (linguistic rules give descriptions pro rata as opposed to numeric values).

Conclusion: Why neuro-fuzzy models?

Neural networks are essentially a 'black box'. A trained ANN system can be tested, and its accuracy can be assured at some level of statistical significance, but the network does not provide any explanation of the problem, whose data it is mapping. ANN systems can provide precise, non-linear correlation between their input and output data, but the mechanism underlying that correlation is opaque. The network parameters (i.e. weights, learning rules, transfer functions, topology, etc.) reveal nothing than can rationally be interpreted as a causal explanation of the real world relationship modelled by that trained network.

This opacity problem has two effects on ANN technology. Firstly, it reduces confidence in ANN technology. Secondly, it makes the design of ANN systems ad-hoc based. Another problem with ANN systems is that with nominal or ordinal representation of input and output useful information could be disregarded. However, this problem of opacity can be overcome by combining ANN systems with qualitative causal models. The most common approach to combining qualitative causal models with ANN systems is the neurofuzzy approach (Zadeh, 1994). Fuzzy concepts can be used to help in understanding the interval information and in combining subjective expertise to generate crisp numbers. Combining neural network systems with fuzzy models helps to explain their behaviour and to validate their performance. Neuro-fuzzy models are fundamentally different from neural and expert systems. Neuro-fuzzy systems have the following characteristics:

- Automatically extract the consequents and the antecedents of a set of fuzzy rules from the original input/output data sets.
- Automatically train and change the shape of member functions according to data patterns.
- The number of neurons are determined from the number of membership functions on each input variable.

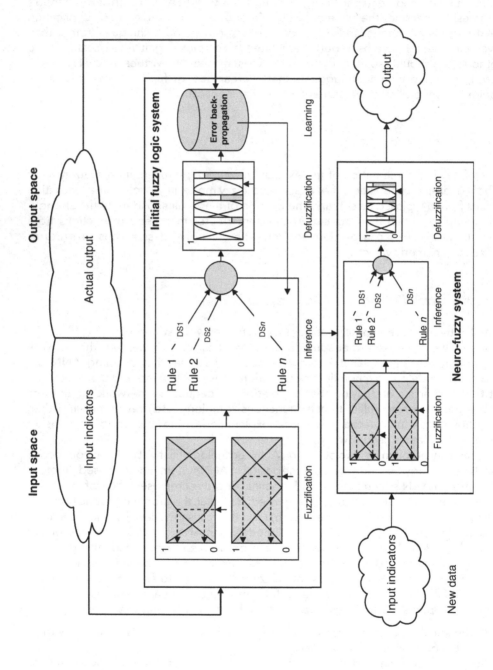

Figure 14.7 Integrating neural networks with fuzzy logic.

- Training and optimisation periods are shorter.
- Allow the inclusion of knowledge and expertise in choosing system topology.
- Lead to a model which can be easily understood.

A generic neuro-fuzzy modelling process is shown in Figure 14.7. The figure shows the life cycle process of developing neurofuzzy models. The process starts with the evaluation and assessment of the suitability of AI tools to solve the problem under investigation. The criteria for evaluation can be found in Boussabaine (1996) and Altrock (1996). Having decided that AI methods are the best for solving particular problems, data has to be collected and analysed. The purpose here is to learn more about the nature of the data and the problem under investigation; to solve problems in data; to change the structure of data (levels of granularity); to extract meaningful knowledge and analyse the qualitative variables and quantitative variables. The next important stage is prototyping. This involves the definition of input and output variables and the types of decision that the neurofuzzy system will make. The expected outcome of this stage is a set of rules that maps out the input to the output. These rules can be extracted automatically from data sets or generated manually by human experts. The generated rules are trained using a modified ANN learning algorithm. This system is then optimised, tested, validated and deployed in the real world if the performance is satisfactory. Necessary feedback is then made for future maintenance and development.

Therefore, the following steps can be deployed in developing a neuro-fuzzy system.

(1) Develop a qualitative model that expresses all aspects of the problem under investigation. The model includes verbal statements (or fuzzy rules) as well as mathematical representations.
(2) Design and train the ANN to:
    (a) Specify each parameter value based on the current input.
    (b) Quantify the qualities expressed by words in the model.
    (c) Predict the outcome of projects by generating a set of outputs from the parameters and qualities that the networks have now quantified.

Recent developments in the uses of Neuro-fuzzy within the building life cycle have identified the system to be reliable and be of value to professionals who require systems, which enable them to make more informal decisions on the allocation and management of construction and operational costs. The system tends to generally work better with large data sets whereas ANN can work with smaller samples as well.

References

Altrock, C.V. (1996) *Fuzzy Logic and Neurofuzzy Applications in Business and Finance*, Prentice-Hall, New Jersey.

Arditi, D. and Pulket, T. (2005) Predicting the outcome of construction litigation using boosted decision trees, *Journal of Computing in Civil Engineering*, 19(4), 387–393, October 2005, *ASCE Journal of Construction Management and Engineering*.

Barron, A.R. and Barron, R. (1989) Statistical learning networks: A unifying view, in Wegman, E.J., Gantz, D.I. and Miller, J.J. (eds) *Computing Science and Statistics: Proceedings of the 20th Symposium on the Interface*, American Statistical Association, Alexandria, Virginia, USA, pp. 192–202.

Boussabaine, A.H. (1996) The use of artificial ANNs in construction management: A review, *Construction Management & Economics, Taylor & Francis Journals*, 14(5), 427–436.

Boussabaine, A.H. (2001) Neurofuzzy modelling of construction projects' duration I: Principles, *Engineering, Construction and Architectural Management*, 8(2), 104–113.

Boussabaine, A.H., Kirkham, R.J. and Grew, R.G. (1999) Estimating the cost of energy usage in sport centres: A comparative modelling approach, in Hughes, W. (ed.) *15th Annual Association of Researchers in Construction Management Conference*, 15–17 September 1999, Liverpool John Moores University, Vol. 2, pp. 481–488, ISBN 0-9534161-2-7.

Chao, L. and Skibniewski, M.J. (1994) Estimating construction productivity: Neural-network-based approach, *ASCE Journal of Computing in Civil Engineering*, 8(2), 841–847.

Demir, Y. and Uçar, A. (2003) Modelling and simulation with neural and fuzzy-neural networks of switched circuits, *The International Journal for Computation and Mathematics in Electrical and Electronic Engineering*, 22(2), 253–272.

Elhag, T.M.S. and Boussabaine, A.H. (2001) Tender price estimation using ANNs I: Data preprocessing, *Journal of Financial Management of Property and Construction*, 6(3), 193–208.

Fausett, L. (1994) *Fundamentals of Neural Networks: Architecture, Algorithms and Applications*, Prentice-Hall International, Englewood Cliffs, NJ.

Flood, I. (2006) Next generation artificial neural networks for civil engineering, *ASCE Journal of Computing in Civil Engineering*, 20(5), 305–307.

Flood, I. and Kartam, N. (1994a) Neural networks in civil engineering I: Principles and understanding, *ASCE Journal of Computing in Civil Engineering*, 8(2), 131–148.

Flood, I. and Kartam, N. (1994b) Neural networks in civil engineering II: Principles and systems and application, *ASCE Journal of Computing in Civil Engineering*, 8(2), 149–162.

Flood, I., Abi-Shdid, C., Issa, R. and Kartam, N. (2007) Developments in coarse-grain modeling of transient heat-flow in buildings, *Journal of Computing in Civil Engineering*, 21(5), 379–382, September/October 2007.

Hall, C. (1992) Neural net technology ready for prime time? *Institute of Electrical and Electronic Engineers (IEEE), Expert*, December, 2–4.

Haykin, S. (1994) *Neural Networks: A Comprehensive Foundation*, Macmillan Publishing, New York.

Huang, M., Tsoua, Y. and Leea, S. (2006) Integrating fuzzy data mining and fuzzy artificial ANNs for discovering implicit knowledge, *Knowledge-Based Systems*, 19(6), 396–403.

Hush, P.R. and Horne, B.G. (1993) Progress in supervised neural networks, *IEEE Signal Processing Journal*, January, 8–39.

Kamarthi, S., Sanvido, V. and Kumara, R. (1992) Neuroform – neural network system for vertical formwork selection, *ASCE Journal of Computing in Civil Engineering*, 6(2), 178–199.

Kohonen, T. (1988) An introduction to neural computing, *Neural Networks*, 1(1), 16.

Kumar, U.A. (2005) Comparison of ANNs and regression analysis: A new insight, *Expert Systems with Applications*, 29(2), 424–430.

Levin, E., Tishby, N. and Solla, S.A. (1990) A statistical approach to learning and generalization in layered neural networks, *Proceedings of the IEEE*, 78(10), 1568–1574.

Lippman, R.P. (1987) An introduction to computing with neural nets, *IEEE, Association Press Journal*, 4(2), 4–22.

Masters, T. (1993) *Practical Neural Network Recipies in C++*, Academic Press, London.

Ok, S. and Sinha, S. (2006) Construction equipment productivity estimation using artificial neural network model, *Construction Management and Economics*, 24(10), 1029–1044.

Rao, M. and Datta, T. (2006) Modal seismic control of building frames by artificial neural network, *Journal of Computing in Civil Engineering*, 20(1), 69–73, January/February 2006.

Rumelhart, D., Widrow, B. and Lehr, A. (1994) The basic ideas in neural networks, *Communication of the ACM*, 37(3), 87–91.

Stergiou, C. and Siganos, D. (2007) Neural networks, accessed at http://www.doc.ic.ac.uk/~nd/surprise_96/journal/vol4/cs11/report.html on 11 April 2008 at 14.56.

Wang, Y. and Elhag, T.M.S. (2007) A comparison of neural network, evidential reasoning and multiple regression analysis in modelling bridge risks, *Expert Systems with Applications*, 32(2), 336–348.

Zadeh, L.A. (1994) Fuzzy logic, neural networks, and soft computing, *Communication of the ACM*, 37(3), 77–84.

Chapter Fifteen

Social network analysis

Stephen Pryke

Introduction

Social network analysis (SNA) is the product of an unlikely collaboration between mathematicians, anthropologists and sociologists. But we should not allow the word 'social' to discourage us from using the method to explore the non-social aspects of the construction process. SNA is essentially a form of structural analysis, allowing mathematical and graphical analysis of what otherwise might be regarded as essentially qualitative data. SNA can be employed to investigate organisational issues as diverse as contractual relationships and communities of practice. The purpose of this chapter is to provide those adopting a 'systems approach' to understanding the construction project team or coalition of firms (Winch, 1989), with some means of useful analysis of such systems. The chapter will outline the limitations within other methods and will provide an introduction to SNA terminology. Some examples of the type of data that might be usefully analysed using SNA will be discussed and there will be a summary of the available software for social network analysts. Finally, a bibliography is provided for those wishing to start using SNA. The chapter draws upon a decade of research involving the largest private and public sector clients in the UK, mainland Europe and The People's Republic of China.

Definition

SNA involves the representation of organisational relationships as a system of nodes or actors linked by precisely classified connections, along with the mathematics that defines the structural characteristics of the relationship between the nodes. Wasserman and Faust (1994) define a social network in even simpler terms as:

> a finite set or sets of actors and the relation or relations (between them).

Research into the activities, and the effectiveness of such activities, in construction-related projects has in the past frequently relied upon what might be regarded as positivist approaches. Although regarded by some as robust, these approaches try to explain what comprises a complex social arrangement (Morris, 1994) through methods that essentially have their origins in natural science (Pryke, 2004). Project management researchers have increasingly looked to the social science disciplines to explain and understand the key issues and problems faced in the management of construction, and non-construction, projects (Bresnen et al., 2005). Some have argued for a need to bridge positivist and interpretivist approaches with more qualitative methods (Chih Lin, 1998), and Loosemore (1998) has argued that SNA is a quantitative tool capable of being applied within an interpretative context in construction research. Loosemore questions the association of quantitative and qualitative methods with causality and the production of universal models, but feels that both quantitative and qualitative methods (jointly) have a part to play in understanding social roles, positions and behaviour in the construction project environment. Others have argued that qualitative and quantitative approaches can be integrated using critical realism (Smyth *et al.*, 2007) and combining extensive and intensive methods is more important than a quantitative–qualitative dichotomy. The main critique of positivism from the critical realist viewpoint is that explanations are both general and particular/context dependent; this position may be appropriate when considering issues associated with the complexity of projects (Smyth and Morris, 2007).

SNA is proposed, therefore, on the basis that it is not used as a method to the exclusion of other methods. Qualitative contextual data is particularly important in the context of construction projects since repeatability is very limited. Speculative housing aside, most construction projects are unique and differ from those that have been completed previously in a number of ways that have impact upon the nature of the roles and activities comprising the project. Having defined SNA, provided a rationale for its consideration and located the method philosophically, it would be useful now to develop an understanding of SNA as a method.

Why choose social network analysis?

Nohria and Eccles (1992) identify five reasons for adopting a network perspective when looking at (not necessarily construction) organisations. These comprise, in summary:

- All organisations operate as social networks; a lot of information and knowledge is transferred through face-to-face interactions, perhaps in official meetings or quite frequently during breaks and at social events.
- Organisations can only operate within a context of other organisations. In other words, an organisation is connected to a network of other organisations, some of which act as clients, others providing services to clients in some form. SNA provides a means of representing and quantifying the relationships that an organisation has with other organisations.
- Organisations are 'suspended in multiple, complex, overlapping webs of relationships and we are unlikely to see … (the whole picture) … from one organisation' (Nohria and Eccles, 1992). Trying to understand how a firm operates and what the issues are in terms of change and improvement to profit and/or efficiency is difficult if the firm is viewed in isolation from its business environment. This outward-looking perspective is most usefully represented as a network of firms and other actors or agencies.
- White (1963) (cited in Nohria and Eccles, 1992, p. 7) sees actors as 'active, purposeful agents who are constantly trying to wrest control for themselves or blocking others from taking control'. Understanding an individual's position in a given network provides a greater understanding of their behaviour and the likely impact of that individual's actions upon the organisation as a whole. The power acquired or asserted by individuals within the organisation that employs them and the temporary organisation within which they operate in fulfilling their role within a project, is an important factor in understanding the effectiveness of organisations.
- The network structure of individual organisations is informative in terms of understanding how one organisation compares with another. The networks (and these might be day-to-day information exchange, or longer-term knowledge transfer networks, for example) provide an insight into the systems operating within each organisation and therefore how such systems might usefully interface with the systems of other organisations to provide a client facing, client satisfying, temporary organisation, in project-based situations.

When we look at construction coalitions, we see a mixture of actor roles being fulfilled by individuals (perhaps the sole trader, professional service provider or subcontractor) and those actors roles fulfilled by a firm; an example of the latter might be the client role, where a firm is named as the client actor, but where daily interactions occur with a number of individuals. The question of how we gather data and analyse it in the context of different actor types is an issue that will be discussed later in this chapter.

Concepts and terminology

A number of social network terms are in common use in the English language. We talk of *webs of relationships, networking, cliques,* individuals who become isolates, or conversely the prominence of individuals, of individuals' centrality and links between firms and individuals; telecoms firms refer to issues of *connectivity* for network users. SNA theory gives very precise meanings to these terms in common use, as well as a large number of other terms, and allocates a formula to each of the terms. It is not the intention to provide SNA formulae here, partly because those wishing to use SNA will want to use one of the numerous software packages available, and partly because Wasserman and Faust (1994) provide what amounts to an encyclopaedia of both terminology and mathematical formulae. It is also worth noting that there is a range of alternative words frequently used in SNA for the same, most common, terms. For example, the words *links, edges, curves* and *connections* are variations on the term that describes the connection between two given nodes or actors. This chapter adheres to the terms used in Wasserman and Faust (1994) to avoid ambiguity. Let us now turn our attention to the definition of the main SNA terms.

Social network

A social network consists of a finite set or sets of actors and the relation or relations defined on them.

Wassermann and Faust (1994, p. 20)

The word 'finite' is important here; so often networks appear to be very extensive, if not infinite, and defining the network boundary can be difficult. Being clear about boundary definition is important. The boundary definition may be achieved by specifying the nature of the role of the actors within the boundary and perhaps time parameters. For example, on one particular construction research project the actors falling within the boundary were classified as those attending the site during a three-month period prior to the interview date, having been identified by another project actor and not using hand tools for all or part of their working day (Pryke, 2001).

Actor

Actors are discrete individual, corporate, or collective social units.

Wassermann and Faust (1994, p. 17)

Networks comprise nodes and connections between those nodes. The node is described as an actor in the network and might be, for example, people in a group, departments within a firm or nations within a world. It is also possible to apply SNA outside of a social context in which case the nodes might be computer terminals or railway stations. For the purposes of studying construction projects, our nodes will normally comprise either individual people or firms. Making the initial decision about

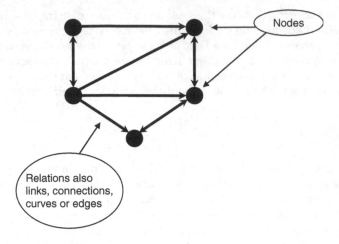

Figure 15.1 Nodes and actors.

whether to measure links between individuals or firms is important, as is consistency of application. If there is homogeneity with the members of the group, we refer to the network as being a one-node network.

Relation

> The collection of ties of a specific kind among members of a group is called a relation. For example … the set of formal diplomatic ties maintained by pairs of nations in the world, are ties that define relations.
>
> Wasserman and Faust (1994, p. 20)

Traditionally in SNA, actors were linked to others by social ties. Increasingly, as SNA research explores new applications, these ties have included links that are not defined as 'social'. In Pryke (2001; Pryke and Smyth, 2006; Pryke, 2006; Pryke and Pearson, 2006) – the definition is expanded to include contractual and financial relationships between firms. Figure 15.1 graphically represents nodes and links (or ties). Some useful examples of the most common types of ties, cited by Wasserman and Faust (1994) are:

- Evaluation of one person by another (expressed friendship, liking or respect).
- Transfer of material resources (e.g., business transactions, lending, or borrowing things).
- Association or affiliation (e.g., jointly attending a social event or belonging to the same social club or networking organisation).
- Behavioural interaction (talking together, sending messages).
- Movement between places or statuses (migration, social or physical mobility).
- Physical connection (a road, river or bridge connecting two points).
- Formal relations (e.g., authority).
- Biological relationship (kinship or descent).

Wasserman and Faust (1994, p. 18)

Examples of application of these types of ties in construction research might include:

- Payments between actors (Pryke, 2001).
- Incentives to perform (Pryke, 2005).

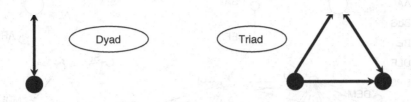

Figure 15.2 Dyads and triads.

- Contractual relationships (Pryke, 2006; Pryke and Smyth, 2006; Loosemore, 1994).
- Instructions issued (Pryke, 2001).
- Information sent and/or received (Pryke, 2004, 2005).
- Risk transferred (Pryke and Ouwerkerk, 2003; Loosemore, 1996).
- Trust (Badi and Pryke, 2006).
- Knowledge transfer (Conway, 1994, 1997).
- Abuse of power and conflict resolution (Loosemore, 1999).

Dyads and triads

Relationships between two or three actors, respectively. A network of only two or three would scarcely comprise a network at all. Dyads and triads tend, therefore, to apply to clusters or subgroups within the main group. Figure 15.2 shows examples of an isolated dyad and an isolated triad.

Subgroup

… we can define a subgroup as any subset of actors, and all ties among them.

Wasserman and Faust (1994, p. 19)

Also known as a clique, a subgroup is a small group within a larger group. Clusters in construction coalitions (Gray, 1996; Holti et al., 2000) and the relationships between those clusters are analogous with Wasserman and Faust's (1994) subgroup. Figure 15.3 gives an example of the representation of clusters; in this case contractual clusters are shown, representing the relationships in traditional construction procurement. The three subgroups clearly evident in this figure represent the client and its advisers, the developer and its consultants and the contractor and its subcontractors.

Network density

… is a concept that deals with the number of links incident with each node in a graph [network].

Wasserman and Faust (1994, p. 101)

Network density is, in effect, an expression of the total number of links between nodes in a given network, expressed in relation to the maximum number of links possible for that network. The maximum number of links possible arises where every node is linked to every other node. When this point is reached the density value is 1.00 and this would represent an unusual situation. Most commonly, the density value would fall between 0.00 and 1.00.

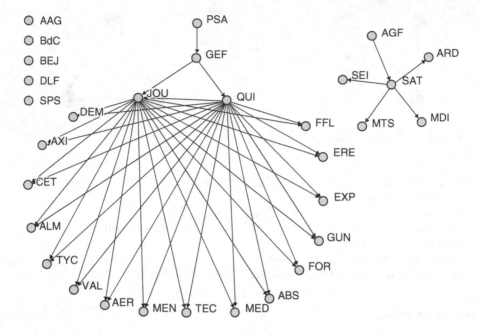

Figure 15.3 Isolates and subgroups. This figure illustrates a number of network characteristics. In the top left of the figure, node references AAG, BdC and so on are isolates – the nodes are not connected to any other nodes; there are no linkages. In the centre of the figure is a network with two very prominent nodes – JOU and QUI; the main network here is not very dense, the majority of nodes are only connected to one or two other nodes and the maximum density of 1 would require each node to be connected to each other node. The small network with SAT at its centre is an isolated subgroup in which SAT has a high level of centrality, this subgroup being quite weakly connected and therefore with a low level of density.

Actor centrality

> Prominent actors are those that are extensively involved in relationships with other actors. We are not particularly concerned with whether this prominence is due to the receiving … or the transmission … of many ties – what is important here is that the actor is simply involved.

> Wasserman and Faust (1994, p. 173)

Centrality of a given actor within a network is an expression of prominence and possibly power, depending on the nature of the relationships being measured. The definition of centrality and its application is complex and a detailed explanation is beyond the scope of this chapter. In simple terms, centrality of an actor refers to the number of links associated with that actor, compared to the total possible number in the context of the network as a whole. At this point the issue of in-degree and out-degree become relevant. In-degree centrality refers to the incoming links; out-degree, the outgoing links. Hence it would be wrong to associate a high level of power within a classic, high centrality, star network configuration, if the centrality is associated with outgoing information or payments, for example, rather than incoming. Figure 15.3 illustrates the point of centrality; the actor at the centre of the star, in very simple terms, has a high level of centrality in the network shown.

Freeman (1979) referred to three main groups of centrality measures: degree of points, betweenness and closeness. The degree of points, or extent to which a given

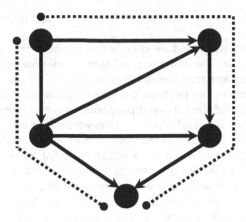

Figure 15.4 Path lengths and bridges. The figure shows two different paths and path lengths between a pair of nodes. The path on the left of the figure involves one bridge, possibly a gatekeeper type of actor positioned between the pair of other actors. The path connecting the pair of nodes and shown on the right of the figure exhibits a longer path length and involves two nodes representing actors fulfilling the gatekeeper function.

point is connected to other points, provides, typically, a measure of communication or interaction activity of some sort. In this case, high degree centrality of a given actor within a network implies a high level of prominence in an information exchange, or some other type of communication network. Point centrality would provide a measure of the importance of an actor, either because the actor was responsible for the very wise dissemination of information (out-degree) or was responsible for gathering information from a large number of other actors (in-degree).

Betweenness centrality relates to the incidence with which a given node falls between two other nodes. Typically an actor with a high value for betweenness centrality has a high level of control over information flowing through them in some way. The actor might typically be acting in a gatekeeper type of role. Finally, closeness centrality, involves the measurement of path lengths between two given points. The concept of path length is illustrated in Figure 15.4. This involves consideration of the number of nodes which fall between two given nodes for, typically, some form of interaction network. This latter measure is perhaps the most complex and gives some measure of the independence of a given actor and the efficiency of the organisation (Freeman, 1979). For the examination of construction coalition networks, degree centrality provides one of the most useful measures for analysis (Pryke, 2001).

Defining the population for the study

Defining the boundary for an SNA study is important. Even a cursory consideration of the mathematics of network density and actor centrality will reveal the importance of making an accurate and appropriate assessment of which actors to include in the network population. It also follows that sampling is not an appropriate approach, unless a very clearly defined subgroup is identifiable. Conversely, in one sense all social networks are huge, transitory and potentially infinitive. Network theorists quote Lauman, Marsden and Prensky's (1989) two possible approaches:

Realist approach to boundary specification – the actors define the boundary of the network themselves. For example, if we start to interview construction coalition actors, they will identify other actors with whom they need to interact to achieve the project objectives.

Nominalist approach to boundary specification – here the network boundary is defined by the researcher. Wassermann and Faust (1994) give an example of the study of computer messages amongst researchers in a given scientific activity. The list is constructed by the researcher, perhaps based on published academic papers in the relevant scientific area and the list is not prepared with any reference to the views of those on the list, particularly in relation to others who might also be on the list.

For the purposes of construction research, we should consider the nature of the network before deciding on whether to adopt a realist or nominalist approach. If the researcher wishes to investigate contractual relationships or information exchange relationships, a nominalist approach would be helpful; contractual conditions tend to be dyadic and the parties to the contract are identified. With information exchange, we would want to consider outgoing and incoming information and if a project coalition member was not identified by either senders or receivers of information, then the actor concerned does not form part of the network. If a nominalist approach is adopted in construction research, there is a tendency towards higher levels of isolates and very often the existence of these isolates is instructive.

Finally on SNA theory and techniques

SNA exists at two quite distinct levels of abstraction. There are the social network theorists who identify social networks as an interesting social construct and explore the implications for society of such networks. Conversely, there is a group academics that wish to understand the mathematical structure of networks and who generally adopt a very broad range of definitions for their networks. In other words, we can analyse the networks of contractual relationships in construction using SNA, even though there is really very little 'social' aspect to such contractual relationships.

Space has permitted only a fairly brief overview of SNA in this chapter and the emphasis has been on the technique, rather than the sociological aspects. For further reading on the sociological aspects of social networks a good start might be made with Scott (2000). For those interested in the mathematical formulae, Wassermann and Faust (1994) provide an encyclopaedic overview of formulae and social network measures.

SNA has a great deal to offer the construction project researcher. The large number of possible variables associated with unique buildings in unique geographical and social settings, means that description and classification of the context for projects that are being compared is important methodologically. An understanding of the project details and the profiles of the actors will provide a richness to the SNA data analysis, the absence of which would otherwise reduce the value of the SNA analysis.

Software for the analysis of networks

Wassermann and Faust (1994) provide mathematical formulae for all SNA measures and examination of these formulae is instructive. Indeed for calculations of network density and actor centrality in networks with fewer than perhaps 50 nodes, the use of a simple calculator is possible and perhaps desirable. Time spent exploring and analysing some simple network data will be rewarded with a better understanding of

Package	Use	Visualisation available?	Support		
			Available[a]	Manual	Help
FATCAT	Contextual analysis	No	Free[b]	No	Yes
GRADAP	Graph analysis	No	Com[b]	Yes	No
JUNG	Modelling, analysis and visualisation	Yes	Free	Yes	Yes
KliqFinder	Cohesive subgroups	Yes	—	Yes	No
MatMan	Matrix analysis	No	Com	No	Yes
NetMiner II	Visual analysis	Yes	Com[c]	Yes	Yes
Pajek	Large data set visualisation	Yes	Free	No	Yes
PermaNet	Permutation tests	No	Free	No	Yes
SNA	R routines	Yes	Free	Yes	Yes
Statnet	Analysis and visualisation	Yes	Free	Yes	Yes
StOCNET	Statistical analysis	No	Free	Yes	Yes
UCINET	Comprehensive package	Yes	Com[c]	Yes	Yes

[a] Free means freeware or shareware; com means commercially available.
[b] This is a DOS package that is no longer updated.
[c] Demonstration copy available on website – see Table 15.2.

Source: Adapted and updated from Huisman and Van Duijin (2003).

Table 15.1 Overview of selected SNA software packages.

the nature of network data and the value of the basic analysis measures. The more experienced analysts and those dealing with large data sets will want to identify some appropriate software packages. Many of these packages have the very important benefit of providing a simple means of generating network graphics, as well as the usual matrices and values.

Table 15.1 provides an overview of some of the more popular SNA software packages available. The table provides a summary of the uses or functions available, whether or not the software will produce some sort of diagrammatical representation of the data and some information about the availability and support for the software. The URL addresses for each of these packages and the home page of the International Network for Social Network Analysis (INSNA) are provided in Table 15.2. The INSNA website is a most valuable resource for any prospective researcher in SNA. In addition to a list of further software packages (although this part of the website is not always completely updated) there are a number of links to publications and current topics within the subject area. Those considering the use of SNA techniques in construction should investigate the possibility of attending the annual conference of INSNA; the research conference details are given on the INSNA website and the conference attracts very large numbers of research active individuals. The conference is attended by the most prominent academics and practitioners in the SNA field. Beginners are always enthusiastically welcomed and abstracts are accepted for research at a preliminary stage of development.

Package	URL
FATCAT	http://www.sfu.ca/~richards/Pages/fatcat.htm
GRADAP	http://assess.com/xcart/product.php?productid=229&cat=32&page=1
JUNG	http://jung.sourceforge.net/index.html
KliqFinder	http://www.msu.edu/~kenfrank/software.htm#KliqueFinder_
MatMan	http://www.noldus.com/
MultiNet	http://www.sfu.ca/~richards/Multinet/Pages/
NetMiner II	http://www.netminer.com/NetMiner/home_01.jsp
Pajek	http://vlado.fmf.uni-lj.si/pub/networks/pajek/default.htm
PermaNet	http://www.meijigakuin.ac.jp/~rtsuji/en/software.html
SNA	http://erzuli.ss.uci.edu/R.stuff/
Statnet	http://csde.washington.edu/statnet/
StOCNET	http://stat.gamma.rug.nl/stocnet/
UCINET	http://www.analytictech.com/
INSNA	http://insna.org [links to software from INSNA site]

Source: Adapted and updated from Huisman and Van Duijin (2003).

Table 15.2 URLs of selected SNA software packages.

Conclusion

This chapter sets out to demystify a research method which has huge potential for application within the field of construction research. Social network analysis enables the researcher to deal with a wide variety of variables using one common method; offering the possibility of more effective comparative study in a field where it could be argued that each project is unique.

SNA provides the construction researcher with a rigorous analytic method to deal with a number of the issues confronting an increasingly complex industry associated with increasingly complex projects. Whether we are exploring information flows, knowledge management structures, the operation of risk transfer or contractual hierarchies, SNA as a method provides analysis to a level of accuracy not available to the construction industry researcher previously. SNA involves no assumptions about hierarchy, which is important and informative when trying to understand how things work in organisations (in our case temporary organisations). The relatively recent and infrequent adoption of the SNA approach is attributable in some part to the complexity of the method. In the past, the barriers to entry for the average student of construction or research practitioner have discouraged many from committing to a study based on SNA.

Before embarking upon a piece of SNA-based research it is recommended that some thought be given to the issues listed above under the heading 'Why choose SNA?' A decision should be made at an early stage as to whether to study networks of individuals or networks of firms. If in doubt, gather data relating to individuals; the data can always be aggregated manually to produce data relating to firms after data gathering is complete. Disaggregating data relating to firms to produce inter-personal relationships is generally not possible and requires a fresh start with data gathering;

at this point you may have lost the goodwill established with the industrial collaborator and you do a disservice to all those researchers trying to convince firms to collaborate in research projects!

It is hoped that this chapter will encourage more individuals to invest the time in making a start with SNA as a method – the study of actors and their relationships, coupled with a very precise means of classification of actor and network attributes, provides huge potential for construction research.

References

Badi, S. and Pryke, S.D. (2006) The effect of trust on the accuracy of design development information flows in UK construction new procurement systems: A Social Network Analysis approach, *Sunbelt XXVI, the Annual Conference of the International Network of Social Network Analysts*, Vancouver, B.C., April 2006.

Barnes, J.A. (1969) Graph theory and social networks, *Sociology*, 3, 215–232.

Bresnen, M., Swan, J. and Goussevskaia, A. (guest editors) (2005) Managing projects as complex social settings, *Building Research and Information*, 33(6).

Chih Lin, A. (1998) Bridging positivist and interpretivist approaches to qualitative methods, *Policy Studies Journal*, 26(1), 162–180.

Conway, S. (1994) Informal boundary spanning links and networks in successful technological innovation, PhD thesis, Aston University, Birmingham.

Conway, S. (1997) Strategic personal links in successful innovation: Link-pins, bridges and liaisons, *Creativity and Innovation Management*, 6(4), Blackwell Publishers Ltd., Oxford, pp. 226–233.

Freeman, L.C. (1979) Centrality in social concepts: Conceptual clarification, *Social Networks*, 1, 215–239.

Gray, C. (1996) *Value for Money*, Reading Construction Forum and The Reading Production Engineering Group, Berkshire.

Holti, R., Nicolini, D. and Smalley, M. (2000) *Building Down Barriers: The Handbook of Supply Chain Management – The Essentials*, CIRIA, London.

Huisman, M. and van Duijn, M.A.J. (2003) *Software for Social Network Analysis*, paper produced by the University of Groningen supported and funded by the Social Science Research Council of the Netherlands Organisation for Scientific Research (NWO).

Laumann, E.O., Marsden, P.V. and Prensky, D. (1989) The boundary specification problem in network analysis, in Freeman, L.C., White, D.R. and Romney, A.K. (eds) *Research Methods in Social Network Analysis*, George Mason University Press, Fairfax/Virginia, pp. 61–87.

Loosemore, M. (1994) Dealing with unexpected problems – do contracts help? : A comparison of the NEC and JCT 80 forms, *Engineering, Construction and Architectural Management*, 1/2, 115–137.

Loosemore, M. (1996) *Crisis* management in building projects – a longitudinal investigation of communication and behaviour patterns within a grounded theory framework, PhD Thesis, University of Reading.

Loosemore, M. (1998) Social network analysis using a quantitative tool within an interpretative context to explore the management of construction crises, *Engineering, Construction and Architectural Management*, 5/4, 315–326.

Loosemore, M. (1999) Responsibility, power and construction conflict, *Construction Management and Economics*, 17, 699–709.

Morris, P.W.G. (1994) *The Management of Projects*, Thomas Telford, London.

Nohria, N. and Eccles, R.G. (eds) (1992) *Networks and Organizations*, Harvard University Press, Boston, MA, USA.

Pryke, S.D. (2001) UK construction in transition: developing a social network analysis approach to the evaluation of new procurement and management strategies, PhD Thesis, The Bartlett School of Graduate Studies, University College London.

Pryke, S.D. (2004) Analytical methods in construction procurement and management: A critical review, *Journal of Construction Procurement*, 10(1), 49–67.

Pryke, S.D. (2005) Towards a social network theory of project governance, *Construction Management and Economics*, 23(9), 927–939.

Pryke, S.D. (2006) Legal issues associated with emergent actor roles in innovative UK procurement: Prime Contracting case study, *Journal of Professional Issues in Engineering Education and Practice Special Legal Section on Legal Aspects of Relational Contracting*, 132(1), 67–76.

Pryke, S.D. and Ouwerkerk, E. (2003) Post completion risk transfer audits: an analytical risk management tool using social network analysis, *Construction and Building Research Conference of the RICS Research Foundation (COBRA 2003)*, University of Wolverhampton.

Pryke, S.D. and Pearson, S. (2006) Project governance; case studies on financial incentives, *Building Research and Information*, 34(6), 534–545.

Pryke, S.D. and Smyth, H. (2006) *The Management of Projects: A Relationship Approach*, Blackwell, London.

Scott, J. (2000) *Social Network Analysis: A Handbook*, Sage, London.

Smyth, H.J. and Morris, P.W.G. (2007) An epistemological evaluation of research into projects and their management: Methodological issues, *International Journal of Project Management*, 4, 423–436.

Smyth, H.J., Morris, P.W.G. and Kelsey, J.M. (2007) Critical realism and the management of projects: Epistemology for understanding value creation in the face of uncertainty, *EURAM 2007*, Paris, May 16–19th.

Wasserman, S. and Faust, K. (1994) *Social Network Analysis: Method and Application*, Cambridge University Press, Cambridge, UK.

Wasserman, S. and Galaskiewicz, J. (eds) (1994) *Advances in Social Network Analysis*, Sage, Beverly Hills, CA, USA.

White, H.C. (1963) *An Anatomy of Kinship*, Prentice-Hall, Englewood Cliffs, NJ, USA.

Winch, G.M. (1989) The construction firm and the construction project: A transaction cost approach, *Construction Management and Economics*, 1989(7), 331–345.

Chapter Sixteen

Managing the thesis

Alan Griffith and Paul Watson

Introduction

The thesis is *the* research product – the culmination of weeks and months and even years of study into a focused and defined subject-specific topic. The evolution of a thesis follows a process and one which is unique to the researcher. At the heart of this process is the development of a research project exhibiting clearly defined aims, objectives, goals and research methods, the ability to critically evaluate literature, to creatively gather information, to accurately analyse and interpret data, to produce robust findings and conclusions and above all else the capability to reflect all of this in a written work amenable to scrutiny and examination. It is a process which is extremely challenging, onerous and dynamic and one which must be actively managed – it does not happen by itself. Although supervisors play an important part in developing and supporting the process, it is the researcher who must drive the process and in so doing lead the job of managing the thesis. To manage the process effectively the researcher needs to understand the process and tasks involved in developing, delivering, assessing and examining a thesis.

Defining the thesis

A thesis has certain unique features, Murray (2002) noted that a thesis is different from any other kind of academic writing. It is the product of years of research and forms the basis of assessment for the award of the degree. It is as advocated by Cryer (1996) the equivalent of scaling an unclimbed peak. Such a difficult task requires commitment and motivation on the part of the researcher, after all as suggested by Delamont *et al.* (1997), if the researcher is not interested in the subject then they are unlikely to complete their thesis. Thus one very important point when writing a thesis is to have an interesting topic upon which to conduct research.

The starting point for writing a thesis is the production of a thesis statement. The thesis statement will assist the researcher in focusing their search for relevant information. A thesis statement is a good starting point because it can act as a means for testing ideas by distilling them into one or two sentences, thus adding clarity of purpose. It can also assist in developing the thesis argument and providing a guide to your argument for the reader. The research project will likely begin with a working, preliminary thesis statement that the researcher will continue to refine until they are certain where the evidence leads.

Postgraduate research of this type is different from undergraduate dissertation writing and usually involves working in isolation (Delamont *et al.*, 1997). It has been argued by Dunleavy (2003) that a thesis is a 'big book' were the author develops and communicates a question and then sets about providing answers. He further advocates that the thesis will report the discovery of new facts, or display the exercise of independent critical power, although these are not mutually exclusive. Phillips and Pugh (2005) corroborate the above but add further that the thinking that links one idea with others and drives the critical thought process has to be unambiguously translated into the written format. One must be careful to avoid what Dunleavy (2003) has identified, that in publishing circles a PhD thesis is often taken to mean an unreadable argumentative, pompous and excessively complex expression of ideas supported by an overkill of referencing. Therefore, to avoid the above, a thesis author should consider what constitutes a good thesis. A good thesis will have a contestable stance, proposing an arguable point, with which other people could disagree. It is also provocative; taking a certain stance and then offering an argument that justifies the discussion presented. Researchers must develop the skill of presenting a logical argument. (Murray, 2002).

Part of thesis production involves surveying the information and views already in existence, which requires the application of 'critical thinking'. Critical thinking is a broad concept that requires the researcher to maintain objectivity and ask questions as they read or gather information. Dunleavy (2003) informs us that independent critical thinking enables the thesis author to demonstrate originality, by presenting a theoretical or thematic argument in an ordered and coherent way. Further, the researcher will have explored already analysed issues from some distinctive perspective of their own. Critical thinking or analysis is an intellectual disciplined activity that combines elements of research, knowledge of historical context, and balanced judgement (Vaughn, 2007). Research students find this aspect of thesis production to be a most problematic activity. However, it is a vital activity in the production of an acceptable thesis and one that must be deployed continually (Rudestam and Newton, 2001). Phillips and Pugh (2005) state that the researcher has to be able to be critical of their own work, as though it was the work of someone else. Setting this distance between you and your work will aid the concept of critical thinking application.

At this point it is worth noting that writing a thesis within the sciences and engineering disciplines is different from one in the humanities and social sciences. In the sciences and engineering, the structure of writing follows very closely the process of the conducted research. The link between research activities and the writing up of the activities is very clear and the whole process is reasonably well structured. Also in many cases the research question is more likely to be developed in conjunction with the supervisor, with research questions being more clearly defined and structured (Murray, 2002). Delamont et al. (2003) note further that in the sciences and engineering the relationship between researcher and supervisor is more of a marriage of interests, with the focus being provided by the supervisor. In the humanities and social sciences it is more likely that researchers will have to invent not only their research question and thesis structure but also find the writing style appropriate for their project. The structure and format of a humanities and social science thesis usually takes a very different form from one in the sciences and engineering disciplines. There is no one best thesis approach in social science, but rather that the approach most effective for the resolution of a given problem depends on a large number of variables, not least the nature of the problem itself (Gill and Johnson, 1997).

Having a clear research focus

For advanced students, the undertaking of a research project can be a daunting and demanding challenge. However, students have to understand that the work is a research project, and should be managed efficiently and effectively as a project. Thus, if the researcher can grasp the concept that their research project is in fact an extension of project management, it becomes possible to address the common threats by the application of some simple project management concepts and tools.

Project management requires a clear focus, as does a research project, so it is essential to start with clearly defined and agreed objectives for the project, though these may be refined later. This point is supported by Rudestam and Newton (2001), who advocate that the prelude to conducting a thesis study is presenting a thesis proposal; a proposal is an action plan that justifies and describes the proposed study.

A researcher has to be concerned with the efficient and effective use of their available resources in the attainment of the aims and objectives contained within a specific plan. This plan may take the form of a research methodology, the plan being the method that needs to be deployed in order to achieve/complete the pre-determined aims and objectives. Researchers determine the broad lines of operation, the strategy or general programme and choose the appropriate methods for the most effective and efficient actions. So planning relates to how, when and where research is to be carried out (Delamont et al., 1997).

However, having a plan is not an end in itself; the plan is only a starting point in trying to control the research project. A successful researcher has also to engage in the control function. After all, as noted by Cryer (1996) and Phillips and Pugh (2005), detailed plans inevitably need regular amendment.

Therefore, let us consider control as applied within the framework of a research project. Control is exercised by the feedback and feed forward of information upon actual performance when compared with the pre-determined plan; therefore planning and control are very closely linked. Control is the activity which measures deviations from planned activities/objectives and further initiates effective and efficient corrective actions based upon a valid comparative analysis.

Developing and managing the draft thesis

Outline structure

An appropriate outline structure to a research thesis typically takes the following form:

- Title page.
- Abstract.
- Acknowledgements.
- Glossary of abbreviations.
- List of contents.
- List of tables.
- Introductory chapter.

Structure of the thesis, research aim(s), objectives, hypotheses, overview of research methodological approach.

- Main chapters.

Background and conceptual development; critical review and evaluation of literature; detailed methodological approach; primary data collection mechanisms;

data acquisition; synthesis and triangulation of data groups; analysis, findings, conclusions, discussion and recommendations.

- References.
- Bibliography.
- Appendices.

What this outline structure actually achieves, in the mind of the researcher, is the transition from a mere conceptual perspective to the actual and detailed research and writing activities which need to be undertaken. This has two real benefits: first, it helps the researcher to identify the key parts of the work which, when combined, will form the whole work; second, it helps in maintaining focus on those key parts as they are undertaken. Structuring the thesis can take many forms and the above is just one example. The structure advocated by Phillips and Pugh (2005) is a well-recognised and popular approach which is not dissimilar to that outlined above. The important point is that the structure sub-divides the whole into parts – the task of sectioning the work.

Sectioning the work

In addition to providing the skeleton for the draft thesis, the outline structure, importantly, sections the work. This is essential in dividing up the total task into manageable pieces which makes undertaking the overall task easier. It is important to remember that the researcher does not, ordinarily, write in a linear sequence from the title page to the appendices, but rather writes in discrete sections which will be ordered appropriately later in the total process of writing.

Planning effort and time

The development of an outline structure into sections is extremely helpful in planning the distribution of effort and assessing the time required to carry out the writing of the draft thesis. Sectioning the work presents stage milestones to the writer, allowing a structured opportunity to pause and reflect on any work stage completed before progressing to the next. Some stages of the writing will require more effort and time than others. A structure, by itself, would not reflect this, but creating defined work stages through sectioning helps the researcher to consider the magnitude of the task for each stage within the project. It is not easy to determine how much time will be needed for each stage. Advanced researchers may have previous experience upon which to determine appropriate timeframes but where this is not the case then advice should be sought from supervisors and from the experiences of other researchers. In any event, having a set project timeframe and a distribution of this into stage durations will help in avoiding overrun.

Managing supervision

Whilst the term 'supervision' implies the direction and inspection of the draft thesis by an over-viewer as it progresses, in practice the researcher takes the lead in directing its development and facilitating its inspection. The researcher really drives the supervision process and therefore, the researcher must 'manage the supervision' of the written work. Appreciating the production of the draft thesis as a set of project stages presents natural milestones within the project where there can be pauses whilst sections of written work can be discussed with supervisors. Each stage of work (for example, a particular written section) should be delivered within the predetermined timeframe. In managing supervision, the researcher should agree with the supervisors at the end of each work stage that: draft work will be presented; review will be undertaken; feedback presented; conclusions drawn; planning of the next stage will be carried out. In addition, it allows the researcher to use these natural pauses to reflect more broadly on the status and progress of the project.

Commitment to writing

One of the common shortcomings displayed by researchers is not committing thoughts to paper. In the early stages of development the researcher will have so many thoughts in their mind, and playing on their mind, that important information can easily be lost. This can be avoided if notes are kept on key aspects as they develop and any associated references listed. This should be done in addition to giving conscious commitment to compiling the draft chapters and actually writing. Some of the early writing will likely require re-writing, editing, cutting and shuffling, while some may be aborted. However, it is far better to have written text which can be edited rather than the substance of the research still in the mind of the researcher. It is best to approach this commitment by confirming that all writing, in whatever form, will be productive, than fear writing because it might be unproductive. Optimising the productiveness of writing is all about maintaining focus.

Maintaining focus

Ensure that the writing maintains a productive edge by co-ordinating the draft chapters as they develop within the outline structure of the work and its sections. As draft text develops the researcher must continuously refer back to the respective section within the structure and identified sub-headings within the section. This should ensure, as far as is practicable, that the text being developed is indeed the text required for the particular section. If the developing work appears appropriate then writing should continue. If it does not then work should be halted and the mind re-focused on the structure, the section and sub-headings until the orientation of the developing text is consistent with the requirements. This may appear to be an obvious matter yet it is easy for a researcher to be so engrossed in the task of writing that the focus drifts away into superfluous areas.

It is also helpful if the task of referencing is conducted as the draft text is written. Again, as text emerges with the train of thought so the references being used should be clear and relevant. It is tempting to come back to referencing at a later stage but in reality this can create difficulty in remembering the references, co-ordination of text with reference lists and sourcing material if it was not listed. Maintaining focus is therefore concerned with constant checking of developing text against the structure and referencing the text as it is written.

Checklist for continuity

Following the tenet that constant focusing, and re-focusing if required, is essential to developing applicable draft writing, systematic checks should be made to ensure the detail of any section is appropriate and to ensure continuity throughout the entire work. This can be undertaken with the aid of a simple checklist. Before any section of work is written, a list of all pertinent content should be compiled, consistent with the structure and, chapter details. As writing is completed within a section and under a sub-heading this can be read, reviewed and ticked off against the checklist. If there are outstanding items to which the researcher must return at a later stage then notes can be made on the checklist as a future reminder to further attention. This should be undertaken before the next section is commenced. Again, this is a simple discipline but one which may save much time later in the process.

The whole draft should also be checked for continuity. This will involve checking the completed draft against the outline structure. If the systematic checks made for each section have been effective then there should be little discrepancy in the final draft check. However, it does provide a further opportunity to act as reassurance that the content of the draft meets that envisaged at the outset and that important aspects have not been omitted. Note that the continuity check has not addressed aspects of grammar, syntax and spelling as these can be checked later. The checks for continuity are concerned with ensuring the correct content and substance has been included.

Producing the final version of the thesis

Reflecting on and modifying the draft

Once the draft version of the thesis is produced a vast proportion of the work will have been completed. However, the version at this stage is not the same version that will be handed-in as the final document. Considerable modification may be required and this starts with the task of reflection. In fact, a 'period of reflection' works best in practice. Over a matter of days, or possibly even a few weeks if time allows, the researcher should review the work, re-appraising its content, style and presentation. Where appropriate, the work should be modified in the interests of clarity, fact and accuracy. As the purpose of this stage is to accommodate current thoughts on content and substance, reference might usefully be made to the continuity checklist of the draft thesis.

Accommodating assessment criteria and marking scheme

Almost all theses which are a component of a taught programme, often termed dissertations, are marked against assessment criteria and employ a marking scheme. Both the assessment criteria and marking scheme will have been communicated to the researcher at the outset, perhaps in the form of explanatory notes. It is essential that the researcher has a clear understanding of these elements as the supervisors' marking of the thesis will be undertaken in direct consultation with these. This is to ensure that the academic criteria specified for the work has been fulfilled and to ensure that the marking of the work is systematic, transparent and robust. If a particular criterion such as 'ability to demonstrate appropriate research methodological approach' is specified with 20 per cent marks allocated to this and the candidate does not address this then a criterion is unfulfilled and marks lost. It is self-evident that the researcher should understand the criteria for the thesis, the marks attendant to these criteria and to ensure that all the criteria are addressed within the thesis. This can be achieved by simply checking the list of criteria and marking allocations in the thesis guide against the content of the thesis. Particular attention should be given to those criteria which have multiple elements as it is easy to address partial rather than whole criteria. A typical example of this is the thesis chapter 'Conclusions and Recommendations' where candidates always address the conclusions but sometimes fail to make recommendations. More often than not the researcher will be provided with a specification of criteria and marking distribution in a single pro-forma document and this can be used to conduct an easy and useful check. For research degree theses the criteria for assessment may be less quantitative and more qualitative. However, these will be clearly specified in the Research Degree Regulations for the relevent university. Additionally, the examiners will be assessing the work to set criteria through a detailed process of questioning at an oral examination – the viva voce – which is described subsequently.

Proof reading

Proof reading is the practice of ensuring that the written version of a thesis can be read and understood by 'the reader'. It should never be assumed that the reader will understand the work simply because the writer does. The writer needs to aid the process of reading, assimilating the content and understanding its message and this can be undertaken quite easily. This involves the writer reading the work and checking the grammar, punctuation and spelling to ensure that the work makes sense in the context of the language used. Moreover, it involves reading the narrative parts in association with any other materials used such as diagrams, illustrations, tables, photographs and referenced works to ensure that these are identified and explained. A further useful method is to let another person read the work to see how easily and clearly they receive the messages in the thesis. The important point with proof reading is that if and when any difficulties arise in any aspect of the thesis then it should be amended and re-checked. It should be noted that proof reading can be quite a long and exhausting task but is an essential contributor to assuring the quality of the final output.

Presentation and final check

Once proof reading has been completed, thought can be given to its presentation. The format and binding of the thesis will likely be specified and these requirements should be adhered to. In addition, it is likely that multiple copies will be required with, usually, a minimum of two copies handed-in. The original copy therefore should remain with the researcher and used as the base copy from which additional copies can be acquired. This base copy should undergo a final check to ensure that all pages from start to end are in the correct and numbered sequence. As photocopies of the thesis are made a final check of each copy should be carried out. It is possible that pages could be moved out of numeric sequence during the copying process. Once this has been done appropriate binding of the loose leaf copies can be undertaken to create a hard bound document. Again, upon receipt of the bound copies a final check should be made to ensure that no error during the binding process has taken place. Once completed, the thesis may be formally handed-in for assessment and marking. It is worth noting that whilst the hand-in milestone marks the end of the development and production stages it is not the end of the research process or the end of the project. The researcher's thought should now be given to an oral examination, where required, or what is termed the viva voce.

Knowing the thesis and preparing for the viva voce

What is the viva voce?

The viva voce is an oral examination. It is conducted by the same examiners who have the responsibility for assessing and marking the written submission. Frequently, an internal examiner will attend together with the researcher's supervisors. Together, external and internal members form a panel before which the candidate sits. Whilst the internal examiner is involved in the assessment process and therefore the event, the supervisor does not actively take part in the discussion but often records notes for later reference. This is particularly important for the provision of accurate feedback to the candidate. The oral examination usually takes the form of a 'discussion' between the examiners and the candidate which focuses on the subject and content of the written work together with wider issues of the potential implications and impacts of the work upon the subject area or discipline. The examiners are seeking to assess the candidate's knowledge of the subject together with a wider understanding. Because the oral examination can address a wide range of issues, the candidate must consciously prepare for the event.

Purpose, approach and format

The viva voce has a number of purposes, the key one's being to: ensure the authenticity of the work (both the research and the written submission); test the candidate's knowledge base of the subject; check that the research has demonstrated both appropriate breadth and depth of investigation; examine the efficacy of the research methodological approach; and determine that the candidate has come to correct conclusions based on the work undertaken and methods used.

The approach used by a panel of examiners can be many and varied. They may adopt a formal and highly structured regime, progressing through the thesis section by section or even page by page and asking very detailed questions. Conversely they may adopt an informal approach, acquiring the reassurance they require through what may appear disparate questioning and discussion. Essentially the approach to the viva voce will be based on the examiners perception of the candidate's understanding as presented through the written work, together with the issues and questions which arise from the examiners reading of the work.

At a Masters level the examination may be up to one hour whilst at Doctorate level the meeting typically lasts between two and four hours. In both situations the meetings can be longer where the examiners require more time to satisfy themselves of the candidate's knowledge and abilities. One of the examiners usually takes the lead role and commences with outlining to the candidate the purpose, approach and format of the meeting. This is accompanied by explaining the method of assessment, marking and modes of feedback to be followed after the meeting. All of this is an introduction to the viva voce and is undertaken to ensure that all persons present have a clear, transparent and complete understanding of the event and its aftermath. The detailed scrutiny of the thesis will then follow, involving review, questioning and discussion of the written work. Members of the team will likely take turns in question-leading, focusing on specific sections of the work together with exploring those issues which emerge from the discussions.

It is important to note that, in most experiences, the examiners come to the viva voce to assist rather than hinder the candidate's potential success. In view of this, the candidate can do much to help the examiners and thereby him/herself. This is where viva voce preparation is necessary and the following activities may prove helpful:

- Re-read the entire thesis to ensure that it is remembered and understood.
- Continue background reading to ensure that the latest information is known.
- Be able to summarise the work and explain it succinctly within a few sentences.

- Know the referenced material used and be able to explain its relevance.
- As well as recognising excellence in the work, recognise weaknesses.
- Anticipate potential questions.
- Request a mock viva with supervisors or colleagues.

In the viva voce:

- Be confident and authoritative but not arrogant.
- Listen to the questions asked and answer those questions.
- Seek clarification if there is any ambiguity during the discussions.
- Be pleasant and agreeable in manner rather than argumentative.
- Discuss and develop viewpoints if required to do so.
- Be honest and humble but do not be overawed.
- Remember the examiners are there to help you achieve if they can.

Typical questions

The examiners questions can take any form but will invariably seek to ensure that their questions allow them to assess the work by scrutinising the following sections and sub-sections:

- Aims, objectives and hypotheses:
    Were the aims clearly stated.
    What were the objectives.
    Were coherent hypotheses developed where appropriate.
    Are there sensible links between the aims, objectives and hypotheses.

- Research methodology:
    Was an appropriate methodology used.
    Were reasons for selection justified.
    Does the methodology meet the needs and the aims and objectives.
    Is there evidence of deeper consideration to the research question.

- Breadth and depth of subject knowledge:
    Does the content reflect a high understanding of the research question.
    Was appropriate reference material used.
    Was there reference to a variety of sources.
    Was there consideration of the validity of sourced material.
    Is there evidence of critical evaluation.
    Is there an original contribution to the body of knowledge
    (applicable at Ph.D. level) and/or new novel or innovative ideas postulated.

- Collection, synthesis analysis and presentation of data and information:
    How was data sourced and collected.
    Was the data representative.
    How was the data collated, analysed and verified.
    How was the data interpreted.
    Was the data appropriately presented.

- Conclusions and recommendations:
    Are the conclusions supported by the data.
    Have the findings been verified.
    Were areas of uncertainty identified and discussed.
    Have the aims and objectives been satisfied.

- Presentation:
    Was an appropriate referencing system used.
    Were conventions of language adhered to.
    Was there clarity of expression, discussion and illustration.

Conclusions

Within this chapter it has been demonstrated that the process of managing the thesis is dynamic. Moreover, the researcher needs to determine their argument clearly and engage with the concept of critical thinking, employing it throughout the thesis production. The task is that of managing a thesis as a project, one with clearly defined goals and engaging with the control function linked to flexible planning. Ensuring that the thesis has an acceptable structure is also an important consideration and should be part of the planning process. In short, the ability to develop and deliver a successful thesis is all about consciously and actively 'managing the thesis process'.

References

Cryer, P. (1996) *The Research Students Guide to Success*, Open University Press, Maidenhead.

Delamont, S., Atkinson, P. and Parry, O. (1997) *Supervising the PhD: A Guide to Success*, The Society for Research into Higher Education, Open University Press, Maidenhead.

Dunleavy, P. (2003) *Authoring a PhD: How to Plan, Draft, Write and Finish a Doctorial Thesis or Dissertation*, Palgrave Macmillan, Basingstoke.

Gill, J. and Johnson, P. (1997) *Research Methods for Managers*, Paul Chapman Publishing Ltd., London.

Phillips, E.M. and Pugh, D.S. (2005) *How to Get a PhD: A Handbook for Students and their Supervisors* (4th Edition), Open University Press, McGraw-Hill Education.

Rudestam, K.E. and Newton, R.R. (2001) *Surviving Your Dissertation: A Comprehensive Guide to Content and Processes* (2nd Edition), Sage, London.

Murray, R. (2002) *How to Write a Thesis*, Open University Press, McGraw-Hill Education, Maidenhead.

Naoum, S.G. (1998) *Dissertation Research and Writing for Construction Students*, Butterworth-Heinemann, Oxford.

Vaughn, L. (2007) *The Power of Critical Thinking*, Oxford University Press, Oxford.

Further reading

Argyris, C. and Schon, D.A. (1974) *Theory in Practice: Increasing Professional Effectiveness*, Jossey Bass, San Francisco.

Dahlgaord, S.M.P. (2004) Perspectives on learning in a literature review, *European Quality*, 2(1), 033–047.

Easterby-Smith, M., Thorpe, R. and Lowe, A. (1991) *Management Research: An Introduction*, Sage, London.

Fisher, A. (2007) *Critical Thinking an Introduction*, Cambridge University Press, Cambridge.

Gill, J. (1986) Research as action: An experiment in utilising the social sciences, in Heller, F. (ed.) *The Use and Abuse of Social Science*, Sage, London.

Chapter Seventeen

Getting your research published in refereed journals

Will Hughes

Introduction

Research must be published, otherwise it will be lost. The most important papers for a researcher to produce are those published in international refereed journals. Good practice in writing papers is something that can be learned. The editorial process involves sending submitted papers to independent experts in the field, usually anonymously, and their comments inform the editor, who decides whether and how to progress with a paper. Much of this is as obscure to experienced researchers as it is to new one's. With forethought and planning, the success rate of getting submitted papers accepted for publication can be increased. Editors and publishers are generally very keen to help people improve their success rate.

Good journal papers are a pleasure to read but for many people, a pain to produce. For some people, writing is a pleasure. For most academics, it is simply essential as the output of research is as good as lost, unless it is published and disseminated. Indeed, the most important feature of an academic CV is the list of publications. It is not just the quantity that matters, but where they are published. The reason that journal papers are often regarded as the most important type of publication is because of the vetting process through which papers are put. Generally, the more rigorous the vetting process, the higher it is rated by the academic community.

Writing good journal papers

A good paper will form a record of progress in research, adding to our collective understanding of the particular topic. Although there is a preponderance of empirical research in a field like construction management, papers which build theory ought to form part of the literature. Generally, papers should either develop or test some kind of theory. It is not necessary to do both in the same paper, but a paper that does neither will not add to the sum of knowledge and therefore will not fall into the category of a research paper. Such a paper should quickly be rejected from a refereed journal and directed instead to a more general interest magazine. There is a big difference between 'archival research journals' and other kinds of periodical.

One very useful and, sadly, unusual way to contribute to theory is to produce a survey paper (also known as a review paper). These seem few and far between, but a good survey paper will critically review the literature in a particular topic or sub-topic, and place the various contributions in relation to each other, showing the emergence and development of ideas and evaluating the relative strengths and weaknesses of the various strands of enquiry already published. By pulling together all the literature in this way, an enormous contribution to our understanding may be made.

Generally, though, most people in our field produce papers that report the results of empirical research.

Structure and style

Any report of research begins with a review of the relevant body of literature, and those who are uncertain about how to do this should look for guidance in Silverman (2000, p. 12) and in Rudestam and Newton (2000: 60–61). In all papers, a structure is required and the argument should flow from one section to the next. Obviously, clear English should be used throughout and jargon should be avoided. There is a great tendency among researchers to use too much jargon. It has its place. If you are conversing with other experts in your field, then an agreed jargon is an extremely shorthand way of communicating large ideas quickly. However, you have to be clear that there is agreement on jargon (this can be helped by providing an index or table of terms) and that the paper is directed exclusively and specifically to such an audience. The general rule is to stick to clear, basic English.

In terms of overall structure, good papers will move from the general to the particular and begin with the context of the work, move through the statement of the problem being investigated, deal with the empirical and/or analytical aspects of the work, then develop the discussion and draw conclusions based upon what has been covered in the paper, relating these back to the original context of the work. Issues connected with style, structure and presentation are dealt with extensively elsewhere in the literature (e.g., Turk and Kirkman, 1989) and there is no need to reiterate that guidance here, other than to state that the easiest questions can be the most difficult to answer: what have you done, why is it important and how have you gone about it?

Conceptualisation and theoretical basis of the work

There should be a clear statement near the beginning of a paper explaining the problem that the paper seeks to resolve. Authors often mistakenly leave this until half way through the paper, or even omit it altogether.

Any serious piece of research will involve concepts that are specific to the issue being investigated, or to the investigative approach that has been taken. These should be summarised and explained if they are not common within the field of the target audience. This is not just a case of explaining the concepts related to the particular phenomena under investigation, but, more importantly, to identify the methodological basis of the work: to answer the question, 'what kind of research is this?' However, a research paper is not place for 'text-book' explanations. You should report what you have done and what you have discovered, and show how it relates to what we already know. Of course, the nature of the investigation is inevitably connected to some issue of relevance in society, or more specifically in the construction industry, but, while it may seem heretical to some, it is not necessary for a piece of construction management research to be practically relevant to industrial concerns. A piece of research may hold relevance only for other researchers, but that should not detract from our judgement of its value. The aim should be to advance our understanding. This is not necessarily the same as increasing the productivity of an industrial sector or the profitability of a company.

There should be explicit connections to an existing body of knowledge or body of theory, although these may not reside in the literature of construction management. Indeed, it is helpful if there are references to bodies of research and knowledge outside our own 'domain', since ours is not an academic discipline in its own right, with its own research techniques and theories (Hughes, 1999). While there are some emerging strands of theory that are unique to construction management or construction economics, most research in this area builds upon theoretical models developed elsewhere in the social sciences. These connections must be identified in order to make clear where a particular piece of work is rooted and to ensure that we are not simply re-inventing theories and models that are well-known in more mainstream disciplines. Without such connections, we run the risk of consigning our research to an academic backwater. With such connections, we may even be able to influence developments in mainstream thinking. In determining the theoretical basis of a piece of research, it is useful to think about knowledge domains.

These issues are important because progress in our understanding often depends upon our ability to generalise from specific examples. One useful question is 'what is the general class of problem of which your chosen topic is a specific example?'. Understanding this enables some kind of view to be developed about the extent to which findings might be generalised into a wider context. Thus, good papers will begin with what is well-known and move gradually deeper into the less well-known (Latour, 1987, p. 57).

In citing the work of others it is important not to merely drop names in. Many authors write a sentence, and then place a citation at the end of it, indicating that someone else has already made this assertion. But this widespread practice is not helpful. First, it does not tell the reader whether the original author concluded a piece of systematic research with this idea, or, perhaps, merely mentioned it in passing. Relying on a statement as true merely because someone else has already written it is not what the citation of literature is supposed to be about. Better practice would not involve mere name-dropping, but would involve a sentence, or part of a sentence, explaining what the authors did to come to their conclusion. And if they did not carry out research in the development of their ideas, then their ideas are no more or less important than anyone else's, and should be treated as such. Ideas and conclusions that have resulted from careful, replicable research should be cited as such, indicating that you understand what other researchers have done, and whether you are citing good or bad research.

The construction of a scientific argument relies on the steady accretion of analytical and empirical results (Latour, 1987). Mind you, as it stands, this assertion is another example of how we all fall into this trap of name-dropping, instead of citing someone's work properly. The sentence should read: *The construction of a scientific argument relies on the steady accretion of analytical and empirical results, a process that has been very clearly explained by Latour (1987) in his discussion of how scientific arguments are constructed.* Thus, we would expect, when seeing names cited as authority for some idea, to be told what they did to come to their conclusion. It is not a question of being right or wrong, just of letting readers understand the context of the ideas we are citing. If you are interested in a more detailed exposition of this kind of thing, Latour's book is a very readable and interesting account of how to make and develop a scientific argument.

Analytical framework and hypotheses

The analytical framework of a research paper is not always clearly articulated. It should be. The extent to which a particular approach is authoritative is often judged in terms of where it has come from. Connections to the research literature should be expected in the passages describing the analytical framework. When this is done well, it makes clear the credibility of a paper by showing the usefulness of the particular approach, or approaches that precede it.

One perennial problem with research papers in our field is the question of whether there should be hypotheses. They are certainly not a prerequisite for a good research paper. In fact, they may not belong at all. The question about whether there should be hypotheses is, perhaps, a wrong question. Their presence or absence depends upon the methodological stance of the research. It is not intended to enter into the methodological debate here, other than to point out the dangers of not understanding the methodological implications of different approaches to research (see, e.g., Seymour and Rooke, 1995). Given one methodological stance, hypotheses may be irrelevant. Given another, they may be indispensable.

If there are hypotheses, they should be clearly stated. If there are no hypotheses, then this, of itself, is not a problem, but it should be clear whether the paper is a review, a case study, a contribution to theory development or some other type of study. Without clear articulation, the reader stands no chance of determining the value of the contribution. In the presence of hypotheses, the relationships between the main variables should be explicit and reasonable. They should be stated in a way that makes them testable and the results, no matter what they are, interpretable. If the research is not built on hypotheses, the significance of the paper's contribution to the development of theory must be explained.

Research design

There are many methods that can be used to find answers to questions. Some are more suitable than others. In answering certain types of question, one particular method may be very powerful, but the same method may be weak in dealing with other types of question. Therefore, the relevance of the methods of research will be judged in terms of their appropriateness to the nature of the question being asked. Similarly, the sensitivity of the methods must match the needs of the research question. A good paper will make clear the type of research design, perhaps by reference to earlier, similar studies from different regions, different industries or different disciplines.

The research must be focused on an appropriate unit of analysis. It is useful to describe the criteria by which this was chosen, as well as the criteria by which the cases were chosen. For example, the unit of analysis could be a person, a finished building, a project, a firm, an industry or a country. Each would result in an entirely different study from the others. Moreover, cases might be selected from a large number of similar cases, which would imply one kind of approach, or the question might be framed in such a way that there is only one case, implying an entirely different approach. Neither, of itself, is more or less valid than the other. Indeed no judgement can be made about the validity of a piece of research simply by counting the cases or referring to the unit of analysis. Each characteristic depends on the other.

It is always important to address whether the research design isolates what is being measured from other effects, or, at the very least, identifies the inter-relationships between the effect under scrutiny and other effects. If the research design involves the identification of variables, they need to be clearly and reasonably operationalised (i.e. translated into simple descriptions of what is measured and how it is to be measured) and the reliability and validity of the measures should be discussed. Similarly, there will be issues related to the appropriateness of the population for the research question being studied, the sample size used and the extent to which the results can reasonably be generalised on the basis of this particular sample.

Again, not all research is as deterministic as this, but there are traditions in different types of work and if a phenomenological or ethnographic approach is being adopted, then the author should take this stance clearly and confidently and not try to dress it up in hypothetico-deductive clothes! These issues are well-articulated by Johnson and Duberley (2000), who warn against the dangers of not dealing with the epistemological positions that are implicit in different approaches to empirical research.

Results and discussion

Within the research paper, the data or evidence of the field-work must be present in some guise. But there are always limits on the length of papers, whether for conferences or for journals. It is inevitable that the data will not be reported in their entirety through these outlets. Thus, one technique is to describe what the data *is like*, rather than what it is, giving examples. The full record of the data can be maintained elsewhere, perhaps in a departmental library or on the internet, so that the interested reader can interrogate the data further.

In any event, there must be sufficient information within the paper itself for the reader to evaluate whether the data were appropriate for the study and whether the data collection and record keeping were systematic. Similarly, the validity and robustness of the results of the study will depend upon whether the analytical techniques were appropriate and adequately described. Most importantly, there should be reference to accepted procedures for analysis. This helps the reader to understand what kind of tradition there is in the particular kind of analysis and how such research is generally reported.

In assessing how systematic the analysis has been, one of the main ideas is to persuade the reader that if he or she were to have done the same things, then the same conclusions would have been reached (Latour, 1987). Again, it is important that this very statement implies a certain epistemological stance, so the researcher and the reader need to be clear about whether they are working from the same basis in coming to their views about the results and their discussion.

Conclusions of a paper

Conclusions can be the most difficult part of a paper to write, particularly if the context and research design have not been addressed properly in the first place. It is often the case that those who have the greatest difficulty writing conclusions can trace their difficulties to poor research planning. When research is well planned, the conclusions become obvious from the work that has been reported.

No new facts should be introduced in the conclusions. The conclusions of the study should be consistent with the results of the analysis. Where there is no numerical analysis, the conclusions should be consistent with, and follow from, the development of the argument in the paper. One thing that is usually not necessary is to include a further summary of the contents of the paper.

In many cases, conclusions can be bolstered by considering whether there are alternative conclusions that are consistent with the data or arguments that have been presented. Also, it is useful to consider both theoretical and practical implications of the results. If the research has been properly contextualised at the beginning of the paper, the theoretical implications of the reported research can be adequately connected to the literature discussed there.

The limitations of the study should be noted, but only in terms of the parameters of the research and applicability of the findings. Authors sometimes misinterpret the purpose of a section on limitations of the work and attempt to indulge in soul-searching self-criticism, identifying faults in the execution and reporting of their own work. This is simply not required. The section on limitations should make clear that, for example, the conclusions do not apply to all construction activity in all places at all times. The approach taken in the research enables certain generalisations to be made, but what are they?

Conclusions can also be bolstered by including discussion of the evidence for and against the researcher's arguments and making a clear distinction between the data and their interpretation.

Elements of a journal submission

A paper submitted to any journal will consist of a manuscript, surrounded by a lot of data about the paper itself.

Covering letter

While most authors write little more than a note asking for a paper to be considered for publication, some go a little further and see the covering letter as an important part of the submission. Gump (2004), for example, shows that there are many things an author can do to expedite the progress of a paper. An author can make clear which institution hosted the research work, indicating qualifications, job title and so on, or at the very least, some indication of why he or she is authoritative on the topic. As papers are sometimes rejected for being beyond the scope of a journal, and in such circumstances would not even get into the refereeing process, it is important to address the letter to the right person, include the title, and the number of words. One very useful piece of advice that Gump provides is to explain to the editor how this paper relates to the scope of the journal. It is also wise to confirm that this is an original submission, and that is not simultaneously being considered elsewhere. There are ethics associated with submitting papers, and it is important to make it clear that you are aware of them. Simple things, like your full contact address and details are often missing from covering letters. Including them can only help. Donovan (2004) adds a further suggestion for covering letters: suggested referees. An editor can be given very useful guidance by an author who suggests two or three referees. Not that they will necessarily be used, but understanding what kind of expertise is best for reviewing a paper will help an editor choose appropriate referees. Similarly, you might wish to provide details of referees to specifically exclude, either because you know of certain individuals who are simply set against your work, or to whom you are related, or who have worked closely with you in the past on this work. Guidance about referees to choose or to avoid is very helpful for editors.

Authorship

The numbers of authors per paper seem to be growing. Analysing data on published papers since 1983 in our field (Hughes, 2000), I discovered that the number of authors per paper was growing, on the average, and faster than the number of institutions per paper. What this means is that although more people are being identified as authors, they are tending to be within the same institution, rather than from different places. This indicates a growth in multiple authorship, but not in inter-institutional collaboration. One worrying aspect of authorship is the question of whether all the authors actually contributed to the writing of the text in the paper. There are different traditions in different areas of science. For example, in some sciences, the head of the institution, the head of the research team, the technicians who provided the resources to enable the research are all cited as joint authors, even though they may have contributed none of the text. In other areas, only those who directly contributed text would be listed as authors. The latter tends to be the case in construction management, although there are some notable exceptions. Because we operate with different assumptions, there is confusion around this issue, and all authors should clarify who will be listed, and in what sequence, before they begin work on their papers, to avoid divisive and difficult arguments later in the process. One alternative to joint authorship, for someone who is not actually an author, is to include mention of them in the acknowledgements.

Acknowledgements

The acknowledgements are frequently missing in papers. In this section, authors can acknowledge the support of the research funders, the contribution of non-authoring colleagues in their research team, the contribution of data subjects, whether anonymous or named, and the contribution of anonymous referees, whose comments often help authors to produce much better papers than would otherwise have been the case. This is not just simple courtesy, but a clear way of communicating to others that you understand the complex processes of research and publication. No one can get through these processes alone, so I would expect acknowledgements in every paper. Of course, it is important that acknowledgements are not sent to referees with the paper for reviewing. One way for accomplishing this is to put the acknowledgements in a separate file, 'not for reviewing', that can be incorporated into the paper after acceptance, before publication. Many journals would do this as a matter of course.

Abstract

Any research paper is capable of being summarised succinctly. Papers are expected to include an abstract or summary at the beginning, especially in the cases of conferences and journals, but this should be the last thing to be written and, as such, may be the least considered but most important part of any paper!

In summarising the contents of a paper, there are several aspects to be borne in mind. Readers of a journal will read most of the abstracts, but very few will read the full papers. Perhaps 95 per cent of readers will read *only* the abstract. The need for abstracts to be terse often causes difficulty and can taint what is otherwise a perfectly acceptable style of writing. Certain problems are frequently encountered. The abstract should not be a table of contents in prose, neither should it be a mere introduction. It should be informative. Tell the reader what the research was about, how it was undertaken and what was discovered, but not how the paper is organised. The main findings must be summarised. If there are too many of them, then just exemplify them in the abstract. Some journals call for structured abstract, but even a journal that does not go that far will deserve an abstract that contains certain essential elements:

- *Background*: A simple opening sentence or two placing the work in context.
- *Aims*: One or two sentences giving the purpose of the work.
- *Method(s)*: One or two sentences explaining what was done.
- *Results*: One or two sentences indicating the main findings.
- *Conclusions*: One sentence giving the most important consequences of the work.

The worst abstract I ever came across had to be completely re-written as it told me nothing about the paper, even though it was not wrong. In fact, I realised later that this abstract could be used to describe most papers I have ever seen. It includes nearly every error in abstract writing, so here it is:

> This paper discusses research which was undertaken in the author's country. A theoretical framework is developed from a literature search and this is used by the authors as the basis of an analytical model. The researchers collected data within this framework and analysed it according to the precepts laid down by earlier researchers in the field. The data is used to demonstrate that our understanding can be significantly increased and this is discussed in the light of previous work. Conclusions are drawn and it is shown that these may be useful for practitioners.

Keywords

I have encountered many strange practices in the way that authors choose keywords for their papers. I am constantly wondering why anyone would add 'construction' as a key-word in a construction journal. Similarly, keywords like 'project site', 'site operations', 'site practice' are all used inconsistently, where any one of them would suit all three purposes. Another common practice is a noun-phrase that combines more than one concept, for example 'procurement case studies', which really should be two keywords, 'procurement' and 'case studies'; in an alphabetical list, someone searching for 'case studies' would miss entirely a paper entered under 'procurement case studies'. Most people are not particularly worried about keywords, so they would not even look for advice on how to choose them. Quite often, authors' keywords are merely the main words from the title, but choosing keywords in this way seems to miss the point.

Help is at hand in the guise of a British Standard; BS 6529:1984 *Examining documents, determining their subjects and selecting indexing terms*. This is a very useful source for working out the difference between good practice and bad practice. Among other things, reading this standard helped me to understand my discomfort with the wide-spread practice of simply copying the words of the title into the keywords. The title denotes the subject of the document, whereas the keywords provide indexing terms for the concepts that are dealt with in the document. Clearly, there is scope for some words appearing in both the title and the keywords, but not without careful consideration.

How, then, should we choose our keywords? It is best to keep keywords simple by using common words, and by not inventing new words for familiar concepts. In choosing specific keywords, it may be useful to consider choosing words from a series of categories, as follows:

- *Discipline*: For example, economics, architecture, statistics, management, organisation, financial accounting, psychology, social science.
- *Methods*: For example, analytical, grounded theory, case study, interviews, experiment.
- *Phenomenon*: For example, information systems, control systems, quality systems, cost systems, procurement, business process, culture.
- *Data source*: For example, construction sector, civil engineering, property development, commercial building, housebuilding.
- *Location*: For example, town, country, region.
- *Unit of analysis*: For example, industry, profession, construction firm, consultancy firm, construction project, design project, briefing, documentation, tendering, construction, occupation, maintenance, disposal, individual.

Choosing a keyword for the discipline of a piece of research is important, but not obvious. For example, in *Construction Management and Economics*, it is pointless including the word 'construction' as a keyword, because, to the extent that it is relevant to one paper, it would be relevant to them all. But a similar paper in *Journal of Law and Economics* would need it. On the other hand, a database of papers including both journals may need what appear to be obvious keywords, but these would need to be added by the database compiler, as they would not belong in the published version of the paper.

As well as advice about how to choose keywords, it may be apposite to provide some advice about what not to do. For example, authors sometimes use the same noun-phrases consecutively; one recent example was 'careers' and 'career development' both for the same paper, another was 'timber panel delamination' and 'timber panel house construction'. While specificity is a good principle, it appears that it is easy to go too far. Another practice to avoid is the provision of keywords that do not help. Perhaps this arises from teasing words out of a meaningful title; for example, 'deployment challenges', 'theoretical framework', 'definition', words that, out of their context, cease to convey anything at all about the papers to which they have been applied.

In this age of electronic databases, keywords may be thought of as an anachronism, since we now have a wide-ranging facility for free-text searches, which examine every word in a document. However, thinking about the notion of keywords, and trying to find articles about them, was a salutary lesson for me when I was articulating the purpose of keywords. Every journal paper contains keywords, preceded by the sub-heading 'keywords'. I wanted to find articles and papers written about keywords. But, a free-text search of papers in which the search term is 'keywords' simply returns every paper published! Similarly, anyone interested in abstract concepts would find nothing by carrying out a free-text search for 'abstract', since most papers contain this word as a sub-heading. Other examples are words like 'building' and 'construction' which are more likely to occur in articles from biochemistry and botany than from the construction sector. Thus, the great advantage of searching for some well-chosen keywords is that there (should have been) some intellectual effort applied to identifying which concepts are covered in the document, which is a more useful guide to its relevance than which words are used. Thus, the main use of keywords from a paper is as index entries in a collection of papers.

There is no recognised thesaurus of construction management research keywords. However, there is a list of suggested keywords in the ARCOM model paper layout, which is available on the internet.[1] This paper provides a short list of 126 potential keywords, but is probably too short, and somewhat outdated as it was compiled a few years ago. By contrast, a major research project a few years ago was carried out for the Joint Contracts Tribunal with the aim of developing a terminology of roles in construction projects (Hughes and Murdoch, 2001). This provided a structured list of definitions of project stages, activities and roles, and is now used as a basis for drafting standard-form building contracts in the UK. Is there a need for similar exercises in other domains of knowledge within construction management?

Research topics come and go as new ideas are disseminated among the research community. But there seems to be a need for a structured list of concepts that could be revised on a regular basis, to enable all of us to navigate our way through the CM literature. Without such a list, keywords will continue to be a wasted opportunity for the research community. Perhaps this is an item for future discussion somewhere.

Tables and figures

It is interesting just how much time is taken up in the editorial process removing unnecessary ornamentation in tables and figures. The inclusion of excessive graphical 'noise' is probably a result of almost universal access to powerful word-processing and graphics software in computers, all with default styles that are difficult to overcome. The best guidance I have seen about portraying complex and comprehensive data visually is from Tufte (2001), who provides an excellent overview of how to develop clear graphical portrayals of complex data. The best figures and tables contain nothing that does not add to the message. It is important to think about whether a journal publishes colour or monochrome figures. If the latter, then convert your graphics into monochrome before submitting them. The worst figures contain less data points per unit area of paper than a sentence of text would. Pie charts, and histograms with only a few data points are simply a waste of paper. Computer screen dumps are frequently out of place in papers that are not about the user interface of software.

References

Clearly, every paper requires connections into the literature and therefore every journal paper will have some references at the end. It is remarkable how some authors either fail to notice that each journal has a specific style for the sake of consistency between articles, or else they do not see it as their task to format their own references. Worse, many references are incomplete, inaccurate or simply missing. This is just clumsy and does not convince editors and reviewers that the paper is carefully prepared and accurately executed. At the risk of stating the obvious, make sure that you format your references according to the expectations of the journal to which you are submitting your paper.

There is one final aspect about the citation of references. This is the extent and pattern of the list of references. Things to avoid in a research paper are extremely short lists of references, or lists that include only the author's own papers. Both of these imply that the author has not made the connections with other work in the topic, or the antecedents of the research. Excessively long lists convey the idea that the author is merely trying to produce the longest possible list. If references are cited properly, with some contextualising text about the meaning and origin of the ideas being cited, then the proportion of the paper occupied by the list of references will not be excessive.

Footnotes and endnotes

Many publishers prefer authors to avoid footnotes and endnotes. They are not always necessary, and should only be used if there is no other way of qualifying or explaining a point in the text. Generally, they are not needed in a well-structured paper and can usually be re-cast into the body of the text.

Editorial processes

On receiving a new submission, an editor or editorial assistant will check that the paper conforms to the requirements of the journal. The topic will be assessed to determine whether the paper falls within the scope of the journal. Tables and figures may be assessed at this point. The length of the paper, and the inclusion of the elements listed above, will also be checked. Some journals may also check at this point that the paper has not already been published. This process is now much easier with tools on the web such as Google Scholar, which provides bibliographical details of a huge range of literature. Referees will also spot papers that have already been published because, if they are experts in their field and up to date with the literature, they often recognise when a paper re-appears. I have often come across papers that were once conference papers, and they re-surface as journal papers. This is acceptable, provided that the newer version of the paper is substantively different, either because it contains fresh data or new arguments, but certainly different conclusions. No academic publisher is interested in re-publishing material that has already been published, quite apart from the potential for copyright infringement. Less frequently, I have come across plagiarism, a very difficult and emotional issue. Fortunately, it is rare. It is also becoming easier to detect with the increasing availability of various computer tools.

In these days of electronic submission, the need to pack securely the correct number of copies of your paper is reducing in importance. But if you are submitting hard copy, check how many copies are required, and pack them properly! It is remarkable how many authors send 100–200 sheets of paper half way around the world in flimsy packaging that falls apart mid-journey.

Receipt and allocation

Once a paper has got through the initial hurdle of being received in good order, fitting with the scope and containing all of the necessary parts of a paper, referees are allocated. The choice of referees hinges usually on the keywords provided for a paper. As discussed above, this can be something of a hit and miss process, unless an editor intervenes and chooses keywords for a paper that are likely to match with keywords against referees' names in the journal's database. Small journals can proceed without the need for a database, as the editor can use personal knowledge of referees and authors to choose suitable reviewers. A further source of possible referees' names is the list of authors whose work is cited in the submitted paper. In *Construction Management and Economics* we have a growing database, currently standing at about 3000 people. This provides us with a wide choice of referees.

In *Construction Management and Economics* we aim to get four referees' reports for each submitted paper. We have recently moved to a fully on-line submission and review process, which enables us to be much more responsive and timely, but also means that we do not wait for slow or absent referees anywhere near as long as we used to. In order to get four reports on a paper, we may have to request anything between six and eighteen people to review the paper. Some decline, some do not answer, some agree to do it and then find they cannot fit it in. These delays may result in a paper being in refereeing for several months. But the new on-line system has inserted an extra step into the process that seems to move things along much more quickly. Whereas we used to send the whole paper to someone, and ask them to review it, we now only send the abstract and invite them to be a reviewer. This makes the process quicker and more effective.

Decisions on papers

Some reports come back the same day, others take a few weeks. Referees' reports are of variable quality. Some are wonderfully argued critiques of the strengths and weaknesses of a paper, fully referenced, showing how the paper can be improved. Others are quick reactions based on a cursory reading, and these are not that helpful. Indeed, such reviews may even be scrapped, especially if they are impolite. The point is, an editor who deals with referees' reports, is not simply counting up votes for and against publication. It is not intended to be a democratic process. What really matters is the strength of argument. A referee who provides a compelling argument to justify accepting, revising or rejecting a paper is far more persuasive than three who simply indicate their views with no compelling argument. A good editor will always say that the decision on publication is the editor's, not the referees'. It is also important for referees to bear in mind that their task is to comment on the quality of the science, not the style of the English, or typographical errors. It is helpful to list those spotted, but this is not the primary role of a referee.

One indicator of a journal's quality is the rejection rate. Most authors would submit their work first to the journal they perceive to be the best. A higher rejection rate would indicate a journal that many people perceive as a worthy place to publish their work. Not all of it gets through the refereeing process. *Construction Management and Economics* currently rejects around 55 per cent of papers submitted, and this rate appears to be rising. Thus, the majority of papers received are rejected at the first hurdle of refereeing. Papers that are not rejected will usually require some form of revision. If the required revision appears to demand further research and a quite different paper, then the decision will be reject and resubmit. If it requires further research work and a revised paper, then the decision will be for major revision. If the paper only requires revision, but with no need for further research work, then the decision will be for a minor revision.

Re-submission

When a revised or resubmitted paper comes back in, unless the revisions were minor, it will go back to the same referees to get their views on whether the changes they suggested have been carried out adequately. Authors are asked to include a detailed account of how they have responded to each comment of each referee, or why they have decided not to follow any particular suggestions. This document is an important part of the resubmission. If the author feels that a referee has made inappropriate comments, or has misunderstood the science, it is important for the author to argue this case. An editor may decide that the referee was simply wrong, in which case, the referee will not be asked to approve the revisions. All this depends on the strength and clarity of the relative arguments. In some cases, where the science is outside of the editor's own experience, we would ask appropriate editorial board members to comment on the competing arguments and advise us about which way to go, but the decision is still the editor's. Referees may come back with further requirements that arise from the revised paper, or merely minor typographical errors that need fixing.

Production

Accepted papers are sent to the publisher immediately, usually on the day of acceptance. In the case of commercially published journals, this moves the focus of activity into the publisher's office, out of the editorial office. A new process starts. The paper is now checked for structure, style and clarity. Once the production office has checked that the basics are all included, the paper is sent for copy editing. This involves someone who is good at English checking the sense and structure of sentences and paragraphs, and drawing up a list of questions for the author about things that are not clear. Once the author has responded to the copy editor's queries, the paper is then on the home straight for appearing in print. It is next sent to a type-setter, an old fashioned job title that used to refer to the placing of metal type into blocks for printing, but these days refers to a computer process that achieves the same end. The purpose of typesetting is to produce the final version of the paper that will be printed.

A paper may spend some time in copy editing, and authors are often surprised by a sudden and urgent request for responses to copy editors' queries after an inexplicably quiet few weeks or even months. Then several more weeks or months can pass until the paper is finally scheduled for publication.

Publication and dissemination

There are essentially two approaches to making up issues of a journal. Either the papers are chosen to go into a specific issue at the time of invitation, on first submission, or issues are made up based on finished papers, after copy editing. In *Construction Management and Economics* both options are used: the former in the case of special issues with guest editors, the latter in the case of all other issues. When we are making up an issue, the sequence of publication of papers is dictated by the date of first submission. This means that older papers are published first, subject to fitting exactly into the number of pages in an issue, topical and geographical spread, and the fact that to encourage terseness in the style of writing shorter papers tend to get chosen over longer one's, all other things being equal.

Once the combination of papers is chosen for an issue, the editorial needs to be written, and we generally ask authors to propose some notes to help with this, as they are generally better placed than anyone else when it comes to describing their work in simple and general terms. After 'making up' an issue there is then about a month or so for the printing to take place.

On publication, press releases are sometimes distributed to trade press about interesting articles, to try and generate interest in the journal and its authors. Authors also get off-prints of their articles, and a copy of the whole journal in which the article appears. While the copy of the journal is of little use, most of us like to have it. What surprises me is how many people do not seem to know what off-prints are for. These days, they are PDF versions of what was printed, but they used to be properly printed copies of the paper provided free to authors, and extra copies could be purchased. The reason for the practice of free off-prints is to help authors to send their recently published work to their peers, to develop their own network of researchers, but that is another topic (see, for example, Agre, 2005).

Conclusion

The publication of research papers is extremely important for the individual researcher, for his or her institution, for the discipline and for society. There are many misunderstandings and plenty of sloppy practice in this endeavour. With a little care and attention to detail, the success rate of authors can be increased greatly. There is plenty of good advice and literature available on all aspects of writing and publishing papers, and some editors are keen to share their experiences, while others are a little more reticent. Understanding the process can only help researchers to become better authors. There should be no mysteries about the publication process. Understanding the process, and the timescales involved, reveals how important it is to enter into dialogue with editors and to keep several papers in various states of progress at any one time. An active researcher could aim to produce papers at the rate of one every month or two. Some will contain significant results, and these should be sent to the highest-rated journals, others will be more routine papers that can be sent to conferences and magazines. As long as there is a steady flow of output being submitted, after a year or so, there will be steady flow of papers being published. The goal is to get the best research published in the best journals for the maximum research impact. Generally, any editor or publisher is happy to help you to achieve this goal.

Note

1   http://www.arcom.ac.uk/conferences.html

References

Agre, P. (2005) Networking on the network: A guide to professional skills for PhD students, Department of Information Studies, University of California, Los Angeles, http://polaris.gseis.ucla.edu/pagre/network.html (accessed 7 November 2005).

Donovan, S.K. (2004) Writing successful covering letters for unsolicited submissions to academic journals: Comment, *Journal of Scholarly Publishing*, 35, 221–222.

Gump, S.E. (2004) Writing successful covering letters for unsolicited submissions to academic journals, *Journal of Scholarly Publishing*, 35, 92–102.

Hughes, W.P. (1999) Construction research: A field of application, *Australian Institute of Building Papers*, 9, 51–58.

Hughes, W.P. (2000) Trends in multiple authorship of papers in construction management (1983–1999), *ARCOM Newsletter*, 15(3), 4–5.

Hughes, W.P. and Murdoch, J.R. (2001) *Roles in Construction Projects: Analysis and Terminology*, Construction Industry Publications, Birmingham.

Johnson, P. and Duberley, J. (2000) *Understanding Management Research*, Sage, London.

Latour, B. (1987) *Science in Action*, Harvard University Press, Cambridge, MA.

Rudestam, K.E. and Newton, R.R. (2000) *Surviving your Dissertation: A Comprehensive Guide to Content and Process* (2nd Edition), Sage, London.

Seymour, D.E. and Rooke, J. (1995) The culture of the industry and the culture of research, *Construction Management and Economics*, 13(6), 511–523.

Silverman, D. (2000) *Doing Qualitative Research*, Sage, London.

Tufte, E.R. (2001) *The Visual Display of Quantitative Information* (2nd Edition), Graphics Press, Cheshire, CT.

Turk, C. and Kirkman, J. (1989) *Effective Writing; Improving Scientific, Technical and Business Communication* (2nd Edition), Spon, London.

Chapter Eighteen

Researcher attitudes and motivation

David Boyd

Introduction

Research is such an elongated journey, riddled with uncertainty, that it puts tremendous stress on individuals. It is not surprising that motivation and progress vary tremendously and this can induce a desire to give up. Indeed, the stamina to succeed may be one undeclared characteristic that is being tested within research. Without it research may never be completed. Thus, one important and under declared competency that is required, is how to survive the stress of undertaking research.

The significance of the individual researcher in research is somewhat complicated and has particular differences in positivistic or phenomenological research paradigms where the researcher is exterior to the research or part of the observation frame respectively. Thus, the individual is either an accurate instrument or an interpreter of practice. In both cases there is a large amount of work involved and the intellectual aspects challenge the self as much as the topic. Thus in this chapter, we are concentrating on the feelings of the researcher towards their task from the perspective of the researcher.

We must look at four aspects to understand this. Firstly, the physical and mental constitution of you, the researcher, which we will call your *Inner Self*. Secondly, your *Personal Environment*, which involves your lifestyle, family, friends and living accommodation. Thirdly, the *Research Project* that you are undertaking, which is an intellectual exercise that is affected by you. Finally, your *Research Environment*, which includes the institution you are working in and the context of your research. These four areas are presented diagrammatically in Figure 18.1. We will look at each aspect in turn and then discuss how researchers can manage themselves to succeed. The key to this is understanding yourself and your environment and using this environment to help yourself.

Inner self

You are the most important part of this chapter. You are the resource which will make your research happen but also you are a barrier which needs to be overcome or at least managed to make the research happen. Central to understanding about yourself is the notion that people are different. That is, you are different from other people. Understanding something of these differences, and what your characteristics are in particular, will help you to cope with things.

Cooper (2002) discusses how people are different both in the way that they see the world and also in the way they experience it. This is both their personality, which is how

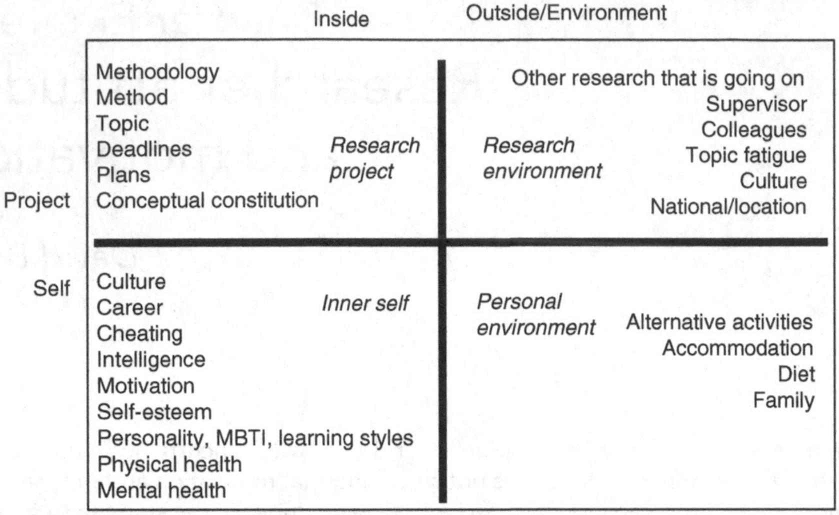

	Inside		Outside/Environment

Figure 18.1   The barriers and drivers of researcher success.

they respond to different situations, and the way they think about events. The latter of these is particularly important in research as it can affect the outcome. This is seldom acknowledged and some research methods try to eliminate the influence of the researcher such that the findings are independent. In a pure science experiment, the researcher is completely independent of the experiment and results require little interpretation by the individual. Conversely, if people are involved as individuals then it is difficult to eliminate the researcher from the research. This difference between people not only is important for you, the individual researcher, and how you respond to situations but for the research project itself, which may involve different people as subjects.

Many differences are subjective (it depends on who is doing the viewing and why) and the significance (what is regarded as being important) of them can be hotly contested. There are lots of models of differences between people which can be useful. For example: Emotional Intelligence (Goleman, 2005), learning styles (Honey and Munford, 1992), Belbin team roles (Belbin, 2000), Multiple Intelligences (Gardner, 1993) and the Myers-Briggs Type Indictor (MBTI) (Myers and McCaulley, 1985). As well as these, it is useful for you to consider your differences concerning motivation (e.g., Pintrich and Schunk, 2002). Each of these has self-administered tests which can allow you to characterise yourself. These tests may be available through your institution or you can find them on-line.

As a researcher, the way that you learn is important as this is how you approach the practice and thinking about any unknown subject. Learning Styles theory, as presented by Honey and Mumford (1992), distinguishes four styles of learning:

(1) *Activists* who prefer having an experience in the here and now and are often gregarious and open-minded but can get bored implementing plans.
(2) *Reflectors* who prefer to stand back in order to gather data and analyse but can be slow at reaching conclusions and may not speak readily.
(3) *Theorists* who prefer to think through events logically seeking to create rational theories.

(4) *Pragmatists* who prefer to find solutions quickly in practical situations but can get bored with meetings discussing decisions.

You might think that the theorist is the learning style you need for research. However, having only this will not enable you to progress and complete the work. On the other hand, if you know you have a particular style you can understand why you find some tasks more difficult and you can put in place strategies and plans for overcoming any problems.

Your motivation for doing research and what keeps you going is important to understand your progress. You need to question why you are doing research. Maslow's theory of needs (Wang, 2001) indicates that the basic physiological and psychological needs, such as food, accommodation and a secure living and working environment, need to be met before higher needs such as excelling at research can be met. Research would be part of self-actualising where it forms part of your identity. However, goals are set to satisfy these needs and these can be long term or short term or involve performance or mastery intentions. Ames (1992) states that performance goals are involved when your abilities are being judged and mastery goals are valued when learning is involved. However, later theories (Pintrich and Schunk, 2002) consider the distinction between external and internal motivations, attribution, self-regulation and self-efficacy. In a sense, there is an expectation that researchers are intrinsically motivated by themselves and that they set their own performance goals. However, because of project and research council pressures, junior researchers will be given performance goals by supervisors. This draws in time management issues (Mackenzie, 1997), which involves identifying a series of short-term goals and monitoring your achievement. Your attitude to this is part of differences in attributional motivation, which considers what you believe are the connections between cause and effect. This provides your explanations for what will happen and why things have happened. Such notions as luck and the weather are external and uncontrollable causes whereas effort and skill are stable and internal causes. It is in this way that we create our self-efficacy which is our judgement of our abilities where clearly a positive belief predicates good performance. Finally, as regards motivation, is our self-regulation or volition. Researchers are expected to be independent learners who possess volition but as Alderman (1999) suggests you need to believe in your abilities to do this and this is formed by developmental factors, socio-cultural factors, attributional history and self-efficacy judgements.

The Myers-Briggs model of personality has four dichotomies: Extrovert and Introvert, Sensing and Intuitive, Thinking and Feeling, and Judging and Perceptive (Myers and McCaulley, 1985). The Extrovert and Introvert dichotomy considers your attitude to life; where if you prefer to direct your energy to deal with people and situations you are an Extrovert, and if you prefer to deal with ideas, information, explanations or beliefs, you are an Introvert. Maybe researchers are more introvert in nature. However, if you are an extreme introvert you might find difficulty in undertaking interviews with different people in their place of work, whereas a questionnaire might be more acceptable. The Sensing and Intuitive dichotomy considers how you perceive, where if you prefer to deal with facts and what you can clearly know you are Sensing, and if you prefer to look into the unknown and generate new possibilities you are Intuitive. Again scientific researchers would tend to be sensing types. However, many aspects of research require intuition about what direction to follow and what theory to generate. The importance of this would arise if you are an intuitive type and find yourself adopting a rigorous experimental approach you may make mistakes and get bored. Contrarily, as a sensing type, you may find it difficult to undertake an observational study requiring empathy with the subjects. The Thinking and Feeling dichotomy considers your decision making where, if you prefer to decide on the basis of objective logic, using an analytic and detached approach, you are a Thinking type, but if you prefer to decide using values and/or personal beliefs, you are a Feeling type. This has many similarities to the Sensing and Intuitive dichotomy. However, it more influences how you might progress and work with others. The last dimension of the Myer-Briggs type indicator – the Judging and Perceptive dichotomy – is important as it considers how you put yourself into the future. That is, if you prefer your life to be planned, stable and organised, you are a Judging type, whereas if you prefer to maintain flexibility and respond to things as they arise, you are a Perceptive type. This can influence heavily the way you prefer to undertake research whether as a structured activity or as the creation of theory from observation. If you find you are using a method (and to some extent methodology) which is contrary to your type, you will find it more challenging and will need to force your practice and thinking to accommodate this.

Emotional intelligence concerns your ability to cope with the situation in which you find yourself. Goleman (2005) places four dimensions on this: self-awareness, self-regulation, self-motivation and social awareness. In many ways, the exercise in this section of understanding your own characteristics and managing the consequences of these is part of self-awareness. In addition, being in control of yourself in order to deliver your research is part of self-regulation. Similarly, the motivational aspects discussed previously, that keep you going through the ups and down of a project, are part of self-motivation. Finally, the fact that you are in an institution, in a research community and in an industry requires you to operate in a social manner; indeed your future success depends on this. Social positioning is dealt with in Belbin (2000) team role classification. This is designed for understanding how differences can fit into a team that has to work and make decisions together. There are nine team roles: Plant, Co-ordinator, Monitor Evaluator, Implementer, Completer Finisher, Resource Investigator, Shaper, Teamworker and Specialist. These consider your willingness and abilities to undertake tasks and how you relate to others while you are doing this.

The foregoing will have given you a flavour of the differences that are identifiable and how the many theories deal with this. There is an idea imposed on you that researchers are clever, capable, dedicated individuals and that what you are doing is excelling at intellectual activity. This strong idea of an imposed norm puts pressure on you, even if it is you that is putting pressure on yourself. The idea that there is a norm puts into question whether you are normal. This is understandable when there are so many differences and it is very easy to see yourself as abnormal or indeed ill. The range of normal is large but your difference is a problem if it affects your ability to work, socialise and live.

At one end of the norm is the question whether you are mentally healthy. The fact that few if any research texts mention individual mental well-being puts even more pressure on individuals to pretend that everything is OK. The stigma behind such problems often prevents people from dealing with them. Mental health involves a wide range of problematic symptoms with medical names such as depression, phobias, and schizophrenia. Many institutions have counsellors and medical staff who can help.

Physical health is also an important and unmentioned aspect of research. Previous accidents and illnesses can leave weaknesses and, like all disabilities, need a plan so that effective adjustment can be made. However, it is general debilitation from colds and other bodily depressors that will cause problems for all people. If illness happens regularly it is worthwhile seeking help. Indeed, keeping a diary may help you to find out whether it may be associated with mental stress resulting from certain activities. The solution is to understand yourself and to find a way to support your needs.

One objective of this section is to build your self-esteem by seeing your particular characteristics as merely differences between people. Self-esteem involves appreciating your own worth and it is important in achieving success through the various problems that will befall any research project. Self-awareness helps you to appreciate the consequences of your differences and self-control makes you do something about it. These models of differences allow you to see how you might respond to the three other areas in Figure 18.1.

Personal environment

In a sense your personal environment is the context that is closest to you and the one that nurtures you. This environment includes people in your friends and family, your accommodation and home, your alternative activities, your diet and your lifestyle. You are to some extent responsible for your personal environment as you put yourself into this situation. As we have indicated, you may have a personality that can accept problems and view them as part of life and your personal environment can be just another part of life. On the other hand, if you are rather melancholy, any problem in your personal environment may add burdens to you personally, and to the project.

The aspects of your personal environment that are in your control are important. There is a good reason to take more attention of your personal environment when you are not under stress, so that, when you are, it does not add further stress and can be part of supporting you. There is a feedback effect that will make matters worse if you fail to recognise the importance of your personal environment. At some point, if neglected, it will require attention and thus put further stress on you when you will have less capacity to cope with it. On the other hand, a supportive personal environment may be able to rescue problems in other areas, thus can support aspects of your personal environment that you are not in control of.

One dimension is about setting a work-life balance. Even though research is more than work, as it is both a passion for discovery and a constituent of your self-identity, continuous study may make you inefficient and ultimately anti-social. Research is not the only thing that can fulfil you and you need to ask what will happen after this research project is complete. Getting and maintaining a job, even in academia, requires a range of skills, one of which is an ability to function in a team and under stress. All but the most brilliant, and in very abstract fields, will not be accepted merely for their ability to undertake thinking tasks. In a sense, if it takes so much out of you to do research that you are debilitated then it may not be sensible persevering.

It is important to have a plan for looking after yourself. This is a matter of taking the resources you have and using them for your survival. This includes accommodation and sustenance. If you are unable to live safely and to sleep peacefully, your body will start breaking down physically or psychologically. An important part of physiological needs is diet. This is as much to do with quality of diet, which affects your whole well being than about quantity. In the long term you are maintaining a biochemical machine and not looking after it or abusing it will cause breakdown. There is much evidence to suggest that a good diet enhances mental abilities.

Exercise is a useful antidote to thinking. Thinking too much can lead one into depression and exercise, for various physiological and psychological reasons, can help to prevent this. An active physical body also helps to keep the biochemical machine working and enhances thinking capacity by stimulating blood flow to the brain. Some exercise activities are also social activities and this can provide an additional support to your life. As well as a poor quality of diet, there are various abuses that make the situation worse. In Britain, the top of this list is alcohol. Whilst this may help you to escape from your problems, it becomes a problem in itself. Other drugs can be more or less debilitating. However, they always have a propensity for addiction, which not only reduces abilities but provides a distraction that needs support both in time and money. The logic of this maintenance of an excellent biochemical machine seldom hits young people, as being young reduces the impact of bad consumption habits. However, once it becomes a necessity for survival it has become a problem and it is crucial to seek help from others.

Another dimension of the work-life balance is your family. Indeed, it may really be your first if you have responsibilities in this respect. Families are not just a distraction but a well of support and encouragement. The economics of being a research novice are not good. This has an impact of what you can do but also the degree of support you can give to a family. Their sacrifice may be a stimulation for you to succeed but it may put extra pressure on you when they have problems such as illnesses or for supporting them in their development such as funding children's school trips.

Other activities can also be regarded as being a distraction, which are acceptable in small amounts. Paid work is a tempting distraction as it helps to support the financing of the body and soul. However, it is another job with the associated pressures of reducing time for relaxation and friends. There may be opportunities for lecturing or tutoring which can use your high level skills and improve your CV. Again the key is balance. If the time and pressures are not too great and the returns are worthwhile, then it is valuable, but this must be part of a plan.

The research project

Having discussed your self and your personal environment then we can start to see how your research project fits into this. The differences between people means that who you are dictates how you see your project. Thus the question of methodology becomes one of 'Do you believe in what you are doing?' and 'To what extent does it relate to your thinking?'. Research being such an all-consuming activity will show up schisms between your beliefs and the beliefs constituted in the project; when this happens your resolve to apply yourself will be tested and your motivation falls. The idea that research is, to some extent, training in multiple methods fails to acknowledge that it is a journey where researchers find themselves and redefine themselves.

Maybe it is your belief that your project is separate from yourself and involves a set of activities set independently where meanings can be discovered with greater accuracy through tasks that are accomplished routinely without disagreement. In this case you will see the constitution of the project as an idealist enterprise. The project then involves determining and enacting good practice which you are a tool of. This may be the case for research into structures and materials, and demographic studies. But, the organisational and individual practice of the construction industry involves a world of fuzziness, conflict and instability where the interaction with the researcher is inevitable.

The formation and progress of the project is important to you. The project itself may merely not interest you or it may challenge your values. Some people are more able to undertake research as a task rather than a vocation but the research may lack inspiration and the application to overcome difficulties. If the research is not well constituted, then progress may be sporadic and difficult to judge. Some research methodologies and methods are based on collecting information of interest about the world and then creating an analysis or interpretation on the results. This can leave researchers feeling extremely vulnerable, with lots of information but nothing to say. The constitution of your project should expect such responses and have strategies for dealing with them. Other methodologies and methods are based on fully structured enquiry with the approach to analysis constituted with the project topic at the beginning. This may appear to deliver a result with certainty. However, the method is very reliant on the assumptions made at the beginning. If these are subsequently found to be unjustified, the whole research project may be worthless. The constitution of such projects needs to anticipate this. A degree of critical reflection on this should be part of the process.

Research, by its nature of being an enquiry into the unknown, involves uncertainty and the likelihood, if not certainty, that events will not transpire without problems. Not only may a research project be populated by problems but the progress may be varied by the nature of the tasks undertaken at different times. A researcher may lose interest in a project. Indeed it is unlikely that they will maintain the level of interest that exists at the beginning. Thus, for a researcher, progress against a plan, which sets up a number of milestone and deadlines, may be important but this may not be achievable because of problem events. It may be that the project is poorly planned, therefore the task that is taken on is not achievable. This is difficult to accept because of a propensity to blame. Either it is your fault for designing it badly or it is your fault for not applying yourself sufficiently to the task. Maybe this is the confidence that more experienced researchers have, that they can constitute projects that are achievable and within this framework they can modify or cope with the circumstances that transpire to deliver results and products, whilst at the same time knowing what it is not possible to do or worthwhile to continue with. When a project is not progressing or if it is not delivering results, then the strength and skills of the researcher are challenged. Sharing the problem with other researchers is a good strategy as it creates an ability to share ideas and to accept feedback, both of which ensure progress.

Research environment

There are two parts to the environment in which the research is taking place. That associated with the institution in which you are working and that associated with the subject of your research.

Institutions have their own agendas, partly set formally and partly set by the social aggregation that makes up the institution. Your immediate environment will include colleagues, supervisors, managers, technical and administrative staff. Individually and collectively they can impact on what you do and feel. This is moderated by the formal processes and wider employment legislation; however, the climate that is set up by those with power and influence often dictates the spirit of this environment. This can set the norm for time commitment, work rate, disclosure, debate and status. For example, it is often contentious that senior people can get credit for work that you have done and also they may get their name mentioned or placed first on a paper merely for their status. This is the norm in academia and institutions should be willing to acknowledge it and reflect on the benefits and problems of such behaviour.

One aspect that may impede progress is the research subject itself in that it can be over-researched. This causes problems of achieving originality but also getting the subjects to take the research seriously. The latter, sometimes called research fatigue, both stops people wanting to get involved and may cause distorted responses due to reacting against the project. The nature of research is that some topics and issues are in-vogue which causes interest in these and even funding behind them. On the other hand, the nature of the built environment world, being heavily time and cash flow driven, makes it sceptical about the value of research. This means that the few organisations and subjects that are interested in research and development are in short supply and indeed are prized possessions. They become the subject of much research and also have to devote time to this. This means that they may become over-researched and may be less willing to engage in further research. This becomes a problem with engagement and also, even if they do engage, this may be at an un-motivated level. In a similar way, populations that receive many questionnaires about their work or perceptions may feel de-motivated to complete them. People receive questionnaires from numerous sources some of which are more legitimate that others, some of which are better structured than others and some of which are part of better constituted projects. However, the respondents are unaware of this and experience mere quantity and see this as an unnecessary distraction. Even if yours is high quality research with a well-constituted purpose it can be very difficult to engage respondents in the work. A very useful perspective in all research is seeing the world from the respondents' point of view. Such problems involving the process of research also lead to de-motivation for the researchers, particularly if they are time constrained.

Another impact of an over-researched topic is that you are likely to be in competition with other researchers. This may be in your own environment or in the wider community. Again, this can be de-motivating as it provides an additional agenda for the research. It is likely to make you and your colleagues secretive. Although this new pressure on time can make you motivated to complete earlier, there is the danger of making mistakes or not perceiving the implications of your study.

The built environment is also an area where research agendas are set outside by government and agencies. This has an impact on the motivation of both the research-ers and the researched. It also has a propensity for influencing the results as the parties need to concur with the external funding agenda. If results do not support a current agenda, then the politics needs to be handled carefully, involving preparing the stakeholders for this revelation. Unfortunately, there is always great pressure to re-interpret results from the dominant position.

The research institution then should be a source of support for the project and for relationships. However, it can be a site of alternative agendas. Working with others in such a contentious environment can be tiring and de-motivating. On the other hand, the life of an institution can drive you to work harder and help to progress work through the climate of generation of ideas. This is the ideal where the institution adds value and the relationships provide personal support. However, there is a danger that an over-social institutional environment can be a displacement activity, which increases mental fatigue by not getting anything done. Institutions and senior researchers need to provide a balanced climate for effective research and enthusiastic researchers.

Conclusion: Keeping going and succeeding

The nature of research, the uncertainty of output, the extended time of activity and the personal challenge means that problems can occur in any of the four areas in Figure 18.1. A researcher's life becomes particularly difficult when they have to deal with problems in more than one area and when these problems start to interact. For example, family illness interacts with institution politics which interacts with priority of topic which interacts with your beliefs about yourself. When things start going wrong, they tend to escalate because of these interactions. In a sense then, in all research projects there is an element of managing yourself and your environment. Part of the problem is that this is seldom acknowledged and so not enough time is spent either dealing with it or, more importantly, planning to cope with it.

The model in Figure 18.1 allows you to work through the interaction of problems. For example, procrastination can be derived from you personally or can occur because the project has particular difficulties. The solution may lie in your personal environment or through systems and pressures in your research environment. Another example is philosophical angst, namely methodological doubt about what you are doing. This can be derived from your inner self in that your beliefs do not align with your supervisors' or the approach of the institution. Again, the solution can be within the research environment itself through finding like-minded researchers for comfort and stimulation or reconstituting the research project itself.

What this chapter has tried to do is to focus on you the researcher. It acknowledges that the researcher is critical to the conceptualisation of the project and to ensuring the progress of it. Problems in the researcher create problem in projects. Importantly, researchers are different and understanding these differences is a critical aspect of doing research. You need to understand what you personally bring to a project and the benefits of this and also the problems. In addition, others are also different and so your family, supervisors and research subjects all see the research differently. It is understanding this and working with it, that delivers successful projects.

- See yourself as an important part of the project within a personal environment and research environment. Look after these.
- Understand your personality and motivation – why you are doing this.
- Recognise what energises you and what drains you. Use this in your task planning.
- Understand the constitution of your project and its implications.
- Plan around the project not just method.
- Set goals that are challenging but achievable.
- Understand the formal and informal institution. How it can support you and how it may be a barrier to your success.
- Learn to discuss problems and accept feedback.
- Plan and monitor major writing. Write from the beginning.
- Balance work and play; body and soul.

References

Alderman, M.K. (1999) *Motivation for Achievement: Possibilities for Teaching and Learning.* New Jersey: Lawrence Erlbaum Associates.

Ames, C. (1992) Classrooms: Goals, structures, and student motivation, *Journal of Educational Psychology*, 84(3), 261–271.

Belbin, R.M. (2000) *Beyond the Team*, Butterworth-Heinemann, Oxford.

Cooper, C. (2002) *Individual Differences* (2nd Edition), Arnold, London.

Gardner, H. (1993) *Frames of Mind: Theory of Multiple Intelligences*, Fontana (Harper Collins), London.

Goleman, D.P. (2005) *Emotional Intelligence* (10th Edition), Bantam Books, London.

Honey, P. and Mumford, A. (1992) *Manual of Learning Styles*, Honey and Mumford Press, Maidenhead.

Mackenzie, R.A. (1997) *Time Trap: The Classical Book on Time Management* (3rd Edition), Amacorn, New York.

Myers, I.B. and McCaulley, M.H. (1985) *Manual: A Guide to the Development and Use of the Myers-Briggs Type Indicator*, Consulting Psychologist Press, Palo Alto.

Pintrich, P. and Schunk, D. (2002) *Motivation in Education: Theory, Research and Applications*, Merrill Prentice Hall, New Jersey.

Wang, S. (2001) Motivation: General overview of theories, in Orey, M. (ed.) *Emerging Perspectives on Learning, Teaching, and Technology*, available online at: http://www.coe.uga.edu/epltt/

Chapter Nineteen

Built environment futures research: The need for foresight and scenario learning

John Ratcliffe

Introduction

The central contention of this chapter is that a more informed, structured and imaginative approach towards the study of the future is demanded of those professions concerned with the stewardship of the built environment and that this can most effectively be provided by the adoption and adaption of the Foresight Principle through Scenario Learning and associated techniques. It explains the work of the *The Futures Academy* at Dublin Institute of Technology, which was established at the millennium with the general aims of:

- *Building* a network of likeminded people from diverse backgrounds who are committed towards the future of business, government and society in Ireland.
- *Identifying* the key drivers of change that might impact on the future of sustainable built environment.
- *Testing* policies directed towards evolving a sustainable future.
- *Providing* executive training and transdisciplinary education in the concepts, methods and techniques of Futures Studies.
- *Developing* a more imaginative approach for decision makers and practitioners involved in forward planning that will ultimately lead to more effective and successful solutions.

Although the remit has been generalised, much of the work to date has related to issues, projects and policies concerned with the built environment. What follows outlines the broad concept of futures studies, the principal methodologies employed and the specific technique of scenario learning which has become a hallmark of the work of The Futures Academy.

Concept and context

Uncertainty has become so pronounced as to render futile, if not downright counterproductive, the kind of strategic planning traditionally employed by governments and corporations – forecasting based on probabilities. Planning for uncertainty increasingly poses the question, 'What is most likely to happen?' (Drucker, 1995). Experience, moreover, has shown us that no unique forecast can be relied upon. Yet, however good our research methods may become, we shall never be able to escape from the ultimate dilemma that all our knowledge is about the past, whilst all our decisions are about the future. As has been averred, moreover, studying futures is not really a question of knowledge and facts at all, but rather one of conjectures (de Jouvenal, 1967). A special approach towards projecting potential futures, so as to improve present decision, is thus required. The 'foresight principle' is such a methodology and 'scenario learning' one of the principal techniques.

Over recent years, it would appear to the author that the dominant tradition of built environment research has been empiricist and retrospective. Considerable effort has been invested in analysing time-series data, and performing ever more elaborate calculations, in order to guide current decision making. Much of the work is derived from other financial markets and social milieu and though improving quality as a comparative and reflective exercise, it can easily overlook many deeper questions – especially those about the future.

What is argued here is that instead of attempting to emulate and employ approaches and techniques drawn increasingly from the quantitative tool-kit of economics, the future study of the built environment should align itself more with qualitative developments in the humanities – in particular, the construction of normative scenarios. 'Normativity' in future studies is considered in slightly different terms from how it is generally considered in the social sciences, where 'norms' are taken as codes of behaviour related to values. In futures studies, normativity indicates the relations of these studies with specific values, desires, wishes or needs or the future.

Taking the general field of futures studies, it has been contended that attempting to impose standards of value neutrality actually runs counter to a paradigm shift that has been taking place in the human sciences (Slaughter, 1996). These are moving away from an understanding of science as requiring the objectivity of the disinterested, value neutral observer. Instead, they have come to accept the inevitability of 'interested engagement' (Ogilvy, 1996). The disinterested, dispassionate 'view from nowhere' (Nagel, 1986) is neither possible to attain, nor appropriate as an aspiration.

Built environment futures

In the context of the built environment, the driving forces of change – cultural, demographic, economic, environmental, governance and technological – impact profoundly upon the way in which we plan, design, finance, develop, occupy, use, manage and value urban form. The world of work is undergoing dramatic changes in respect of who does work and for how long, where it is done, when it is done, how it is done and what we even mean by work. One thing for certain is the workplace of the future will be very different from that of today. Likewise, the house of the future is likely to be more than a home; it will also probably be a family centre for learning, working and taking leisure. The advent of electronic shopping already poses serious questions about the prospective size, location and nature of retail development, as well as the consequent effects upon, and adjustments to, the built environment. Leisure is now a vital component of developed economies, and tourism one of the fastest growing industries worldwide. Economic growth, demographic shifts, technological innovation and social change will all conspire to make the leisure and tourism industries difficult to predict and uncertain to finance and develop. Financial markets, in their turn, are becoming more homogeneous at one level, yet more sophisticated and customised at another. Who can tell what the comparative risks and returns across and between the various classes of the investment market in general, and the sectors of the property market in particular, will be five, ten or fifteen years hence. Traditional methods of research and forecasting – rational, scientific, objective, trend based, lineal and quantitative in nature – are simply not capable of searching, appreciating, testing and explaining the intricate, multivarious yet interrelated imbroglio of issues, factors and players that determine the future shape and performance of the built environment in all its many guises. To be capable of creating the future for the built environment, one has first to be able to imagine it. Futures studies, through foresight, provide the possible answer.

Futures concepts

It is generally accepted that for anything new to be done, there needs to be a rationale, a way of justifying what is intended, and attention focused on some of the expected outcomes. Four key points initially emerge from the concept of futures studies.

(1)   Decisions have long-term consequences.
(2)   Future alternatives imply present choices.
(3)   Forward thinking is preferable to crisis management.
(4)   Further transformations are certain to occur.

The crucial questions most usually facing futures researchers in the examination of an issue or policy include.

- What are the main continuities?
- What are the major trends?
- What are the most important change processes?
- What are the most serious problems?
- What are the new factors 'in the pipeline'?
- What are the main sources of inspiration and hope?

A useful metaphor to describe the aim of the futures field is to provide a 'map of the future'. In essence, futures studies supplies policy makers and others with views, images and alternatives about futures in order to inform and protect decisions in the present. It is important to note, that the underlying purpose of future studies is not to make predictions, but rather to gain an overview of the present human context in order to illuminate alternative futures. Interpretation not forecast.

Moreover, there is a growing realisation, in all areas of life, that the future is not fixed. The notion that the future can be 'shaped' or 'created' has gained currency over the past decade, and is increasingly the basis upon which organisations of all kinds make their plans. As Charles Handy (1989) put it:

The future is not inevitable. We can influence it if we know what we want it to be.

Nothing, the mantra runs, is more certain than the unpredictability of the future. Experience, moreover, demonstrates that no special forecast can be relied upon. What is required, therefore, is not just the hindsight of so much built environment-related research, but also the foresight of the land and urban development professions' leading minds that collectively could inform, influence and inspire impending decision making in sustaining the built environment. We have measured the past and surveyed the present, now we need the confidence to explore the future. The 'foresight principle', enacted through 'foresight programmes', largely in the scientific and tech-nological fields, provides an opportune, seductive and feasible approach for studying the future of the built environment.

Foresight defined

Foresight is a systematic, participatory, future-intelligence-gathering and medium-to-long-term vision-building process aimed at present day decisions and mobilising joint actions (FOREN, 2001). It arises from a convergence of trends and underlying recent developments in the fields of 'policy analysis' and 'strategic planning', as well as futures studies. Foresight brings together key agents of change and various sources of knowledge in order to develop strategic visions and anticipatory intelligence. In this way, there are said to be five essential elements of foresight – anticipation, participation, networking, vision and action (ibid). A particular manifestation of this has been the undertaking of what are known as technology foresight exercises in a number of countries worldwide. Typically, the technology foresight process involves the identification of various principal sectors, the formation of consultative panels for each sector comprising leading figures from the field, and the use of either 'Delphi' or 'scenario planning' techniques, or both, to identify, explore and test future potential markets and prospective opportunities in a climate of risk and uncertainty. From this, robust and flexible strategic policy options are chosen, and a shared mindset among the various participants is fostered. The most important aspects of the Foresight process have been précised as being (Irvine and Martin, 1984):

- Communication between parties concerned.
- Concentration on the longer term.
- Co-ordination of research and development.
- Consensus created on future directions and policies.
- Commitment generated among those responsible for implementation of policy.

In essence, foresight is the process of attempting to broaden the boundaries of perception by carefully scanning the future and clarifying emerging situations. This is done as follows.

- Assessing the implications of present actions and decisions (consequence assessment).
- Detecting and avoiding problems before they occur (early warning and guidance).
- Considering the present implications of possible future events (proactive strategy formulation).
- Envisaging aspects of desired futures (normative scenarios).

A generic approach to describing a comprehensive foresight project has been suggested by Peter Bishop et al. (2007) as in Table 19.1.

A defining characteristic of foresight is that, in essence, it is a human capacity to think ahead and to forecast possible outcomes of present decisions (see Figure 19.1).

	Step	Description	Product
1.	Framing	Scoping the project	Project plan
2.	Scanning	Collecting the information	Information
3.	Forecasting	Describing alternative futures	Scenarios
4.	Visioning	Choosing a preferred future	Prospective
5.	Planning	Organising resources	Strategic plan
6.	Acting	Implementing the plan	Action plan

Table 19.1 Approach to a foresight project.

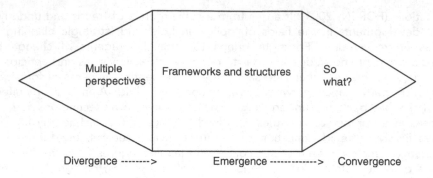

Figure 19.1 Conceptual approach. *Source:* Outsights.

The future function of foresight

It has been asserted that (Slaughter, 1999, p. 169):

> Under modern conditions foresight is less a choice than a necessity with all the force of an historical imperative.

This author would concur, though not perhaps in such hyperbolic or declaratory terms. The world is changing at a pace hitherto unknown, and in directions previously unimagined. There is a need for academic institutions and professional societies in the field of the built environment to play a more positive role in pursuing some or all of the following tasks.

- Raising issues of common concern that may be overlooked taking the traditional short-term view. This might include the provision of decent affordable housing for all, the establishment of stable urban ecosystems and generally building a liveable city!
- Highlighting dangers, alternatives and choices that require attention before they become critical. Global warming is perhaps the most obvious, but the investment in alternative sources of energy and tackling the threat that poverty in cities poses for the built environment are others.
- Publicising the emerging picture of the near-term future in order to involve the public and contribute to present-day decision making. The changing nature of work, urban transport congestion, resource conserving mobility and the need to promote the compact city by densification to combat urban sprawl all spring to mind.
- Contributing to a body of knowledge about foresight and the macro-processes of change and continuity that frame the future. The establishment of a Built Environment Futures Forum would be a welcome first step towards this.
- Identifying the dynamics and policy implications of the transition to sustainability. The most basic principle of all is the need to foster sustainable urban development of all kinds, at all levels, and in all quarters. Easy to state, but difficult to operationalise.
- Helping to identify aspects of a new world order so as to place these on the global political agenda. The most fundamental issue to tackle in urban development and governance is social and political exclusion. Reducing not just poverty as such, but also dependence and isolation. Economic redistribution should go together with some degree of social equity.

- Facilitating the development and application of social innovations. The key in all analysis of social change being first to identify and understand the general trends and forces at play, and to demonstrate their effects in individual cities or other urban areas with their own history, economy, cultures and traditions.
- Helping people to deal with fears and become genuinely empowered to participate in creating the future. Despite the progressive victory of democracy over authoritarianism, two major dilemmas remain – handling pressure group politics and tackling the failure of local democracy in many cities and regions throughout the world.
- Assisting organisations to evolve in appropriate ways. Put simply, the most effective way in which foresight can help is to collect and disseminate 'best practice' in the planning, development and management of the built environment in a relevant, reliable and realistic way.
- Providing institutional shelters for innovative futures work which might not be carried out elsewhere. Here, the universities have a significant role to play in generally promoting the study of futures, and the schools or faculties of the built environment within them, in embodying the art of foresighting as an essential approach in the various disciplines represented.

Returning to Slaughter, he argues that academic institutions should (ibid):

> Play a major part in the process since they have the talented people and much of the expertise required. This suggests that they lift their focus beyond culturally conservative forms of knowledge, short-term politics, boundary maintenance and credentialism.

Prospective

A particularly progressive and proactive form of foresight is to be found in 'prospective'. The prospective, or more familiarly 'la prospective', has French origins, but is now being more popularly applied across Europe in a variety of strategic planning settings. In the francophone context, however, prospective refers to a much wider approach and activity than other futures methodologies as it comprises not only the study of the future, and an evaluation of alternative outcomes against given policy decisions, but also the will to influence the future and to shape it according to society's wishes. Furthermore, it is a very formalised, inclusive, comprehensive and rigorous methodology when compared to more generalised future studies. In many ways, it is similar to foresighting, but would be better understood as a specific means of applying the foresight approach. The two methodologies have been constrasted as: foresight would be the capacity to hear, but prospective would refer to the proficiency to listen to particular things. Put another way, prospective covers the concepts of 'preactivity' (understanding) and 'proactivity' (influencing), whereas foresight concerns itself with 'preactivity', but the idea of 'proactivity' is missing (Godet, 2001). In any event, the term prospective and its application across a broad range of policy issues on a wider territorial basis than hitherto is likely to gain greater currency over the next few years.

Causal layered analysis

Several different techniques have been developed over the past few years which introduce and apply a 'layered' approach towards futures studies, prospective processes and foresight methods. In general, they enable the practitioner to move progressively to ever greater levels of understanding as new strata of meaning are discovered or developed. A generic term for the technique is Causal Layered Analysis (CLA). It has been found to be particularly productive in futures workshops where participants come from different cultures or possess disparate ways of viewing a problem. CLA is best used prior to scenario construction as it permits a vertical dimension for a framework by which scenarios of different types can be created and compared.

Some of the main benefits of CLA have been described as follows (Inayatullah, 2004).

- Expanding the range and enriching the content of scenarios.
- Helping to combine different ways of knowing and understanding among participants.
- Appealing to a wider range of individuals with contrasting ways of thinking and expressing themselves.
- Stratifying participants' positions, both conflicting and harmonious.
- Moving the debate and deliberation beyond the superficial and obvious to the deeper and more cryptic.
- Allowing for an array of transformative actions to be appraised.
- Producing policy options that are informed by alternative layers of analysis.

Commonly, four layers of analysis are pursued. Slaughter (2002) describes them as 'pop'; problem-oriented; critical; and epistemological. Inyatullah (*op cit*) labels them: litany; social causes; structure and discourse; and metaphor and myth. And Voros (2005) characterises his four major 'strata of depth' as: external artefacts or 'constructs of consciousness'; internal artefacts or 'contents of consciousness'; internal processes or 'capacities of consciousness'; and external 'conditions of existence' or 'life conditions'. At The Futures Academy we tend to use a simplified three-layered approach: 'empirical'; 'interpretive'; and 'exploratory'. All, however, are based ultimately on the Delphi injunction 'Know thyself' and the desire to conduct visioning exercises of whatever form with new vistas of insight.

Scenarios

One of the most popular and persuasive techniques used in futures studies and foresighting is scenario analysis. Scenarios have long been used by government planners, corporate strategists and military analysts as powerful tools to aid in decision making in the face of uncertainty. They are instruments for ordering people's perceptions about alternative future environments in which today's decisions might play out. In practice, scenarios resemble a set of stories built around carefully constructed plots. Such stories can express multiple perspectives on complex events, with the scenarios themselves giving meaning to these events.

Despite their story-like qualities, scenarios follow systematic and recognisable phases. The process is highly interactive, intense and imaginative. It begins by isolating the decision to be made, rigorously challenging the mental maps that shape people's perceptions, and hunting and gathering information, often from unorthodox sources. The next steps are more analytical: identifying the driving forces, the predetermined elements and the critical uncertainties. These factors are then prioritised according to importance and uncertainty. Subsequently, three or four thoughtfully composed scenario 'plots', each representing a plausible alternative future, against which policy options can be tested and implications identified, are developed. Then, the deeper structures and systems behind the scenario stories, and their underlying logics, are elaborated to explain them and reveal their crucial differences. Finally, the key events, or turning points, that would channel the future towards one scenario rather than another are identified.

The prime aim of scenarios and scenario building is to enable decision makers to detect and explore all, or as many as possible, alternative futures so as to clarify present actions and subsequent consequences. They should, thus, be prevented from making strategic decisions before they have done some strategic thinking!

According to Michel Godet (1994), scenarios should aim to detect the key variables that emerge from the relationship between the many different factors describing a particular system, especially those relating to the particular actors and their strategies. In doing so, they provide a context for thinking clearly about the otherwise impossible complex array of factors that affect any decision; give a common language to decision makers for talking about these factors and encouraging them to think about a series of 'what-if' stories; help lift the 'blinkers' that limit creativity and resourcefulness; and lead to organisations thinking strategically and continuously learning about key decisions and priorities (Schwartz, 1996). Intrinsically, however, all scenarios must be plausible, structurally different, internally consistent, challenging and useful.

Scenarios are said to deal with the core problems of a given futures study (Coates, 1996). Individual trends do not automatically come together to create useful pictures of the future applicable to planning. A primary purpose of scenario building, therefore, is to create holistic, integrated images of how the future might evolve. These images, in turn, become the context for planning, a testing ground for ideas, or the stimulus for new development. A scenario may further be used to describe a future state, and thereby form the basis for policy analysis. Conversely, the scenario may tell a complete story including the possible or probable policy actions and outcomes. In addition to some future state, scenarios may describe the transition from a present to a future state (Coates, ibid). They can also create alternative histories, describe histories that did not come about, or that would have come about if a certain factor had been altered (Inayatullah, 1996).

Five basic functions of scenarios can be distinguished (Rienstra and Smokers, 1996).

(1) The signalling function; scenarios provide better insight into certain situations.
(2) The communication and learning function; scenarios stimulate thinking about alternative futures.
(3) The legitimation function; scenarios mobilise people and start processes of change.
(4) The exploring and explaining function; scenarios show how solutions for specific problems may become reality, given certain policy priorities, as well as present possible solution strategies.
(5) The demonstration function; scenarios show the consequences of specific decisions.

Ultimately, however, the purpose is not just about constructing scenarios, it is about informing decision makers and influencing, as well as enhancing, decision making. In this context, it has been suggested (Fahey and Randall, 1998) that the purpose of scenario building is to:

- Augment understanding by helping to see what possible futures might look like, how they might come about, and why this might happen.
- Produce new decisions by forcing fresh considerations to surface.
- Reframe existing decisions by providing a new context within which they are taken.
- Identify contingent decisions by exploring what an organisation might do if certain circumstances arise.

In adopting and executing scenario planning and learning for the above purposes every organisation can be seen to face three fundamental challenges. How can it learn from the future before it happens? How can that learning be integrated into decision making? And how can an organisation prepare for the future better and faster than its competitors? Even cities, and development projects within cities, have to compete these days in much the same way as businesses. It has been argued, therefore, that all agencies seeking how to improve how they learn lessons from the future in order to plan today must master a number of specific tasks (ibid).

- Understand the present – and the past.
- Describe a variety of potential futures.
- Delineate how such futures might evolve.
- Identify appropriate indicators to track.
- Link alternative future views to specific decisions.
- Link the scenario process to the key analysis process.
- Link to organisational processes to ensure widespread participation.
- Involve the decision makers.

One of the central roles played by scenario thinking is to provide a kind of 'laboratory' or 'wind tunnel' in which alternative models of the future environment can be tested. Five different uses for scenarios in this context have been described by SRI International as follows (Ringland, 1998).

(1) *Sensitivity/Risk Assessment.* Using scenarios as a wind tunnel for new project proposals or new business developments.
(2) *Strategy Evaluation.* Using scenarios as 'testbeds' to evaluate the viability of an existing strategy.
(3) *Strategy Development: Using a 'Planning Focus'.* Employed within a project or management team to foster a better understanding and build a more robust strategy.
(4) *Strategy Development: Without using a 'Planning Focus'.* This takes a wide range of scenarios at face value and tests strategies against them without judging probabilities.
(5) *Skills.* Using scenarios to reduce fear, uncertainty and doubt and help to formulate training and recruitment needs.

Turning briefly to the purpose of scenario planning and learning in the specific context of the built environment, perhaps the overriding appeal is their capacity for handling complexity. In one of the seminal papers on the use and development of scenarios, a leading luminary, Pierre Wack (1985), writes about the need for 'strategic vision' and the relationship it has with scenario planning in coping with turbulence and uncertainty. Scenarios act as a 'complexity reducer', a common frame of reference within which information can be judged and decision makers helped in discerning what signals to look for against the 'noisy' background of the external environment (ibid).

Above all, scenario planning can create a 'learning organisation'. But that organisation must have the will, the insight and the stamina to undertake such a learning process, as well as making available the resources to make the necessary investment to develop the skills required to construct and employ those scenarios to identify, analyse and manage uncertainty. Good scenarios, moreover, always challenge and surprise – bad one's merely confirm current conceptions and perpetuate personal prejudices.

Types of scenario

Many valid methods of building scenarios exist. At the risk of oversimplification, however, scenario construction can be divided into two basic forms.

(1) *Future backward* – several significant futures are selected and attempts are made to discover the paths that lead to them.
(2) *Future forward* – based on an analysis of present forces and their likely evolution, several sets of plausible futures are projected.

Generalising in the same way, scenarios usually have four dimensions (Inayatullah, 1996).

(1) *'Status Quo'* which assumes that the present will continue into the future. Also known as 'more of the same'.
(2) *'Collapse'* which results when the system cannot sustain continued growth, or when the contradictions of the status quo lead to an internal decay or crash.
(3) *'Steady State'* which is based upon a return to some previous time, either imagined or real, that was perhaps quieter, slower or generally less commercial, industrialised or densely populated.
(4) *'Transformation'* which presumes some fundamental change that may be spiritual, technological, political or economic.

Another way in which scenarios can be categorised has been described by Fahey and Randall (1998 op. cit.) as follows.

- *Global scenarios* which offer leaders a guide to a number of distinctive future environments that each have different implications for long-term investments, operating decisions and options analysis
- *Industry scenarios* which enable managers to identify plausible future states of an industry and differences between them, to examine how these distinct industry states might evolve, and to determine what the organisation would have to do to win within each industrial future
- *Competitor scenarios* which offer a unique method of identifying and testing plausible competitor strategy alternatives in various circumstances
- *Technology scenarios* which help management to make better technological decisions by better understanding the opportunities, risks and choices in preparing for a dynamic, turbulent and uncertain future market.

Yet a further way in which scenarios can be typified is shown in Figure 19.2.

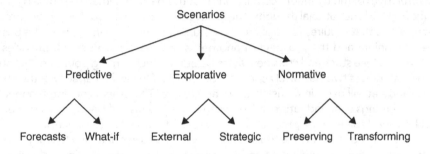

Figure 19.2   Borjeson scenario typology (quoted in Bishop *et al.*, 2007).

Scenario techniques

Scenarios have been categorised by Bishop and others (op. cit.) into eight basic types.

(1) *Judgement*: including genius forecasting, visualisation and role playing.
(2) *Baseline*: producing only one scenario, the expected future, and conducted by trend extrapolation, systems scenarios or trend impact analysis.
(3) *Elaboration of fixed scenarios*: in explicit consideration of multiple scenarios previously prepared using 'incasting' or the popular SRI matrix methods.
(4) *Event sequences*: employing such approaches as probability trees, sociovision or divergence mapping.
(5) *Backcasting*: adopting horizon mission methodology, Impact of Future Technologies or future mapping.
(6) *Dimensions of uncertainty*: using morphological analysis, field anomaly relaxation, GBN matrix or MORPHOL.
(7) *Cross-impact analysis*: calculating the relative probabilities of occurrence of future events and conditions in alternative scenarios employing a computer programme, such as SMIC or IFS.
(8) *Modelling*: having recourse to trend impact analysis, sensitivity analysis or dynamic scenarios.

Prospective through scenarios

Over the past decade the author has developed a futures methodology which he has applied to a wide range of built environment policy fields. It is called 'prospective through scenarios', and the process is shown in Figure 19.3.

A detailed explanation of the process is described elsewhere (Ratcliffe, 2001).

The simple but profound message is that we do not have to walk blindly into thinking and planning the future development of the built environment. There are recognised methods of foresighting drawn from other fields of inquiry which help provide a fuller understanding of the forces shaping the longer-term future of our towns and cities. Some elementary applications are beginning to emerge in both the construction industry and the sustainable cities movement. Indeed, the author's own work on the use of Prospective Through Scenarios is helping to formulate a model which might be applied with benefit elsewhere.

Conclusion

The central myth of the twentieth century was that the path to human destiny is by way of scientific method and rational thought. These have been over-valued, and a more productive path is one that requires foresight and the pursuit of wisdom. In this way, the passing of one millennium and the prospect of another is not merely symbolic; it provides the opportunity to take stock and consider where we stand. Such turning points are important because they reflect two powerful aspects of our reality. One is the capacity of the human mind to range at will over time past, present and future. The other is our interconnectedness with all things past and future. It follows that the promise of the twenty-first century cannot be found solely in the products or processes of rational intelligence, in displays of technological virtuosity or in new tools or techniques. It lies in our ability to learn from the past and strike out in new directions by embracing Futures Studies, adopting the Foresight Principle and applying a methodology such as Prospective Through Scenarios.

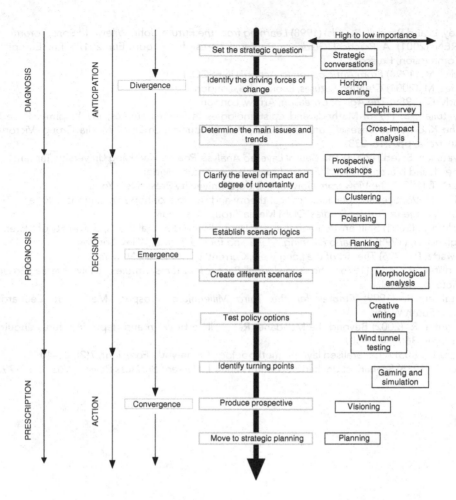

Figure 19.3 'Prospective through Scenarios'.

Perceptively, Albert Einstein once averred something along the lines that no problem can be solved from the same level of consciousness that created it: we must learn to see the world anew; familiarly, he also declaimed that:

> Imagination is more important than knowledge

Thus, it is finally proposed that built environment theory can best be formulated by recourse to futures, foresight and, most of all, imagination.

References

Bishop, P., Hines, A. and Collins, T. (2007) The current state of scenario development, *Foresight*, 9(1), 5–25.

Coates, J. (1996) An overview of futures methods, in Slaughter, R. (ed.) *The Knowledge Base of Futures Studies*, Vol. 1, Foundation DDM Media Group, Victoria, Australia.

de Jouvenal, B. (1967) *The Art of Conjecture*, Weidenfeldt and Nicholson, London.

Drucker, P. (1995) *Managing in a Time of Great Change*, T.T. Dutton, New York.

Fahey, L. and Randall, R. (eds) (1998) *Learning from the Future*, John Wiley and Sons, Toronto.

FOREN (2001) *A Practical Guide to Regional Foresight*, Report Eur 20128 En. European Commission, Brussels.

Godet, M. (1994) *From Anticipation to Action*, UNESCO, Paris.

Godet, M. (2001) *Creating Futures*, Economica, London.

Handy, C. (1989) *The Age of Unreason*, Arrow, London.

Inayatullah, S. (1996) Methods and epistemologies in futures studies, in Slaughter, R. (ed.) *The Knowledge Base of Futures Studies*, Vol. 1, Foundation DDM Media Group, Victoria, Australia, pp. 216–233.

Inayatullah, S. (ed.) (2004) *The Causal Layered Analysis Reader*, Tamkang University, Taiwan.

Irvine, J. and Martin, B. (1984) *Foresight in Science*, Pinter, London.

Nagel, T. (1986) *The View from Nowhere*, Oxford University Press, New York.

Ogilvy, (1996) Scenario planning, critical theory and the role of hope, in Slaughter, R. (ed.) *The Knowledge of Futures Studies*, DDM Media Group, Melbourne.

Ratcliffe, J. (2001) Built environment futures, Unpublished doctoral thesis, University of Ulster.

Ringland, G. (1998) *Scenario Planning: Managing for the Future*, Wiley, London.

Schwartz, P. (1996) *The Art of the Long View*, Currently Doubleday, New York.

Slaughter, R. (ed.) (1996) *The Knowledge Base of Futures Studies*, DDM Media Group, Victoria, Australia.

Slaughter, R. (1999) *Futures for the Third Millennium*, Prospect Media, St. Leonards, New South Wales.

Slaughter, R. (2002) Beyond the Mundane: Reconciling breadth and depth in futures enquiry, *Futures*, 36(8), 493–507.

Voros, J. (2005) A generalised layered methodology framework, *Foresight*, 7(2), 28–40.

Wack, P. (1985) Scenarios: Unchartered waters ahead, *Harvard Business Review*, V63(5), 72–79.

Index